D0408946

RUBBISH!

RUBBISH!

THE ARCHAEOLOGY OF GARBAGE

William Rathje and Cullen Murphy

HarperCollinsPublishers

To the memory of Thomas Price

HarperCollins books may be purchased for educational, business, or sales promotional use. For information, please call or write: Special Markets Department, HarperCollins Publishers, Inc., 10 East 53rd Street, New York, NY 10022. Telephone: (212) 207-7528; Fax: (212) 207-7222.

FIRST EDITION

Designed by Ruth Kolbert

LIBRARY OF CONGRESS CATALOGING-IN-PUBLICATION DATA

Rathje, William

Rubbish! : the archaeology of garbage / William Rathje and Cullen Murphy.—1st ed.

p. cm.

ISBN 0-06-016603-7

1. Garbage Project (University of Arizona) 2. Refuse and refuse disposal—Research—New York (N.Y.) 3. Refuse and refuse disposal—United States—Case studies. I. Murphy, Cullen. II. Title.

TD793.3.R38 1992

363.72′8′097471—dc20 91-50452

---

92 93 94 95 96 DT/RRD 10 9 8 7 6 5 4 3 2 1

CONTENTS

PART I

An Introduction to the Garbage Project

CHAPTER 1

YES, WONDERFUL THINGS

On a crisp October morning not long ago the sun ascended above the Atlantic Ocean and turned its gaze on a team of young researchers as they swarmed over what may be the largest archaeological site in the world. The mound they occupied covers three thousand acres and in places rises more than 155 feet above a low-lying island. Its mass, estimated at 100 million tons, and its volume, estimated at 2.9 billion cubic feet, make it one of the largest man-made structures in North America. And it is known to be a treasure trove—a Pompeii, a Tikal, a Valley of the Kings—of artifacts from the most advanced civilization the planet has ever seen. Overhead sea gulls cackled and cawed, alighting now and then to peck at an artifact or skeptically observe an archaeologist at work. The surrounding landscape still supported quail and duck, but far more noticeable were the dusty, rumbling wagons and tractors of the New York City Department of Sanitation.

The site was the Fresh Kills landfill, on Staten Island, in New York City, a repository of garbage that, when shut down, in the year 2005, will have reached a height of 505 feet above sea level, making it the highest geographic feature along a fifteen-hundred-mile stretch of the Atlantic seaboard running north from Florida all the way to Maine. One sometimes hears that Fresh Kills will have to be closed when it reaches 505 feet so as not to interfere with the approach of aircraft to Newark Airport, in New Jersey, which lies just across the waterway called Arthur Kill. In reality, though, the 505-foot elevation is the result of a series of calculations designed to maximize the landfill's size while avoiding the creation of grades so steep that roads built upon the landfill can't safely be used.

Fresh Kills was originally a vast marshland, a tidal swamp. Robert Moses's plan for the area, in 1948, was to dump enough garbage there to fill the marshland up—a process that would take, according to one estimate, until 1968—and then to develop the site, building houses, attracting light industry, and setting aside open space for recreational use. ("The Fresh Kills landfill project," a 1951 report to Mayor Vincent R. Impelliteri observed, "cannot fail to affect constructively a wide area around it. It is at once practical and idealistic.") Something along these lines may yet happen when Fresh Kills is closed. Until then, however, it is the largest active landfill in the world. It is twenty-five times the size of the Great Pyramid of Khufu at Giza, forty times the size of the Temple of the Sun at Teotihuacan (see Figure 1-A). The volume of Fresh Kills is approaching that of the Great Wall of China, and by one estimate will surpass it at some point in the next few years. It is the sheer physical stature of Fresh Kills in the hulking world of landfills that explains why archaeologists were drawn to the place.

To the archaeologists of the University of Arizona's Garbage Project, which is now entering its twentieth year, landfills represent valuable lodes of information that may, when mined and interpreted, produce valuable insights—insights not into the nature of some past society, of course, but into the nature of our own. Garbage is among humanity's most prodigious physical legacies to those who have yet to be born; if we can come to understand our discards, Garbage Project archaeologists argue, then we will better understand the world in which we live. It is this conviction that prompts Garbage Project researchers to look upon the steaming detritus of daily existence with the same quiet excitement displayed by Howard Carter

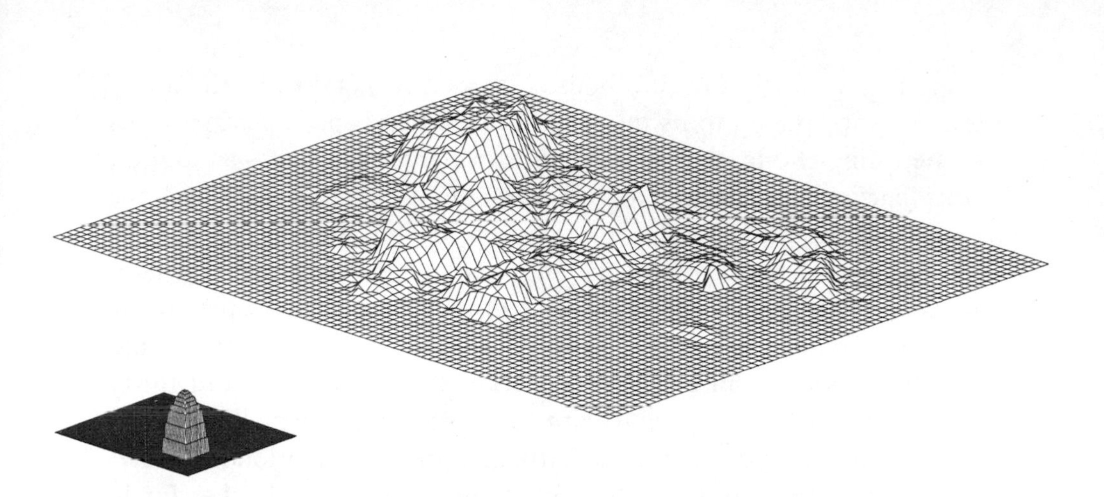

Figure 1-A. A comparison of the Pyramid of the Sun at Teotihuacan, in Mexico (left) and the Fresh Kills landfill, on Staten Island, in New York (right). The Pyramid of the Sun is roughly 800 feet to a side; the Fresh Kills grid as a whole represents an area roughly 2.8 miles by 3.8 miles. Elevations have been exaggerated for clarity, but the relative volumes represented are accurate.

SOURCE: Masakazu Tani, The Garbage Project

and Lord George Edward Carnarvon at the unpillaged, unopened tomb of Tutankhamun.

"Can you see anything?" Carnarvon asked as Carter thrust a lighted candle through a hole into the gloom of the first antechamber. "Yes," Carter replied. "Wonderful things."

Garbage archaeology can be conducted in several ways. At Fresh Kills the method of excavation involved a mobile derrick and a thirteen-hundred-pound bucket auger, the latter of which would be sunk into various parts of the landfill to retrieve samples of garbage from selected strata. At 6:15 a.m. Buddy Kellett of the company Kellett's Well Boring, Inc., which had assisted with several previous Garbage Project landfill digs, drove one of the company's trucks, with derrick and auger collapsed for travel, straight up the steep slope of one of the landfill mounds. Two-thirds of the way up, the Garbage Project crew directed Kellett to a small patch of level ground. Four hydraulic posts were deployed from the stationary vehicle, extending outward to keep it safely moored. Now the derrick was raised. It supported a long metal rod that in turn housed two other metal rods; the apparatus, when pulled to its full length, like a telescope, was capable of penetrating the landfill to a depth of ninety-seven feet—enough at this particular spot to go clear through its bottom and into the original marsh that Fresh Kills had been (or into what was left of it). At the end of the rods was the auger, a large bucket made of high-tension steel: four feet high, three feet in diameter, and open at the bottom like a cookie cutter, with six graphite-and-steel teeth around the bottom's circumference. The bucket would spin at about thirty revolutions per minute and with such force that virtually nothing could impede its descent. At a Garbage Project excavation in Sunnyvale, California, in 1988, one of the first things the bucket hit in the cover dirt a few feet below the surface of the Sunnyvale Landfill was the skeleton of a car. The bucket's teeth snapped the axle, and drilled on.

The digging at Fresh Kills began. Down the whirring bucket plunged. Moments later it returned with a gasp, laden with garbage that, when released, spewed a thin vapor into the chill autumnal air. The smell was pungent, somewhere between sweet and disagreeable. Kellett's rig operator, David Spillers, did his job with the relaxation that comes of familiarity, seemingly oblivious to the harsh grindings and sharp clanks. The rest of the archaeological crew, wearing cloth aprons and heavy rubber gloves, went about their duties with practiced efficiency and considerable speed. They were veteran members of the Garbage Project's A-Team—its landfill-excavating arm—and had been through it all before.

Again a bucketful of garbage rose out of the ground. As soon as it was dumped Masakazu Tani, at the time a Japanese graduate student in anthropology at the University of Arizona (his Ph.D. thesis, recently completed, involves identifying activity areas in ancient sites on the basis of distributions of litter), plunged a thermometer into the warm mass. "Forty-three degrees centigrade," Tani called out. The temperature (equivalent to 109.4 degrees Fahrenheit) was duly logged. The garbage was then given a brusque preliminary examination to determine its generic source and, if possible, its date of origin. In this case the presence of telltale domestic items, and of legible newspapers, made both tasks easy. Gavin Archer, another anthropologist and a research associate of the Garbage Project, made a notation in the running log that he would keep all day long: "Household, circa 1977." Before the next sample was pulled up Douglas Wilson, an anthropologist who specializes in household hazardous waste, stepped up to the auger hole and played out a weighted tape measure, eventually calling out, "Thirty-five feet." As a safety precaution, Wilson, like any other crew member working close to the sunken shaft on depth-measure duty, wore a leather harness tethered to a nearby vehicle. The esophagus created by the bucket auger was just large enough to accept a human being, and anyone slipping untethered a story or two into this narrow, oxygen-starved cavity would die of asphyxiation before any rescue could be attempted.

Most of the bucketfuls of garbage received no more attention than did the load labeled "Household, circa 1977." Some basic data were recorded for tracking purposes, and the garbage was left on a quickly accumulating backdirt pile. But as each of what would finally be fourteen wells grew deeper and deeper, at regular intervals (either every five or every ten feet) samples were taken and preserved for full-dress analysis. On those occasions Wilson Hughes, the methodical and serenely ursine co-director and field supervisor of the Garbage Project, and the man responsible for day-to-day logistics at the Fresh Kills dig, would call out to the bucket operator over the noise of the engine: "We'll take the next bucket." Then Hughes and Wilson would race toward the rig in a running crouch, like medics toward a helicopter, a plywood sampling board between them. Running in behind came a team of microbiologists and civil engineers assembled from the University of Oklahoma, the University of Wisconsin, and Procter & Gamble's environmental laboratory. They brought with them a variety of containers and sealing devices to preserve samples in an oxygen-free environment—an environment that would allow colonies of the anaerobic bacteria that cause most of the biodegradation in landfills (to the extent that biodegradation occurs) to survive for later analysis. Behind the biologists and engineers came other Garbage Project personnel with an assortment of wire mesh screens and saw horses.

Within seconds of the bucket's removal from the ground, the operator maneuvered it directly over the sampling board, and released the contents. The pile was attacked first by Phillip Zack, a civil engineering student from the University of Wisconsin, who, as the temperature was being recorded, directed portions of the material into a variety of airtight conveyances. Then other members of the team moved in—the people who would shovel the steaming refuse atop the wire mesh; the people who would sort and bag whatever didn't go through the mesh; the people who would pour into bags or cannisters or jars whatever did go through the mesh; the people who would label everything for the trip either back to Tucson and the Garbage Project's holding bins or to the laboratories of the various microbiologists. (The shortest trip was to the trailer-laboratory that Procter & Gamble scientists had driven from Cincinnati and parked at the edge of the landfill.) The whole sample-collection process, from dumping to sorting to storing, took no more than twelve minutes. During the Fresh Kills dig it was repeated forty-four times at various places and various depths.

As morning edged toward afternoon the bucket auger began to near the limits of its reach in one of the wells. Down through the first thirty-five feet, a depth that in this well would date back to around 1984, the landfill had been relatively dry. Food waste and yard waste—hot dogs, bread, and grass clippings, for example—were fairly well preserved. Newspapers remained intact and easy to read, their lurid headlines ("Woman Butchered—Ex-Hubby Held") calling to mind a handful of yesterday's tragedies. Beyond thirty-five feet, however, the landfill became increasingly wet, the garbage increasingly unidentifiable. At sixty feet, a stratum in this well containing garbage from the 1940s and 1950s, the bucket grabbed a sample and pulled it toward the surface. The Garbage Project team ran forward with their equipment, positioning themselves underneath. The bucket rose majestically as the operator sat at the controls, shouting something over the noise. As near as anyone can reconstruct it now, he was saying, "You boys might want to back off some, 'cause if this wind hits that bucket. . . ." The operator broke off because the wind did hit that bucket, and the material inside—a gray slime, redolent of putrefaction—thoroughly showered the crew. It would be an exaggeration to suggest that the victims were elated by this development, but their curiosity was certainly piqued, because on only one previous excavation had slime like this turned up in a landfill. What was the stuff made of? How had it come to be?

What did its existence mean? The crew members doggedly collected all the usual samples, plus a few extras bottles of slime for special study. Then they cleaned themselves off.

It would be a blessing if it were possible to study garbage in the abstract, to study garbage without having to handle it physically.* But that is not possible. Garbage is not mathematics. To understand garbage you have to touch it, to feel it, to sort it, to smell it. You have to pick through hundreds of tons of it, counting and weighing all the daily newspapers, the telephone books, the soiled diapers, the foam clamshells that once briefly held hamburgers, the lipstick cylinders coated with grease, the medicine vials still encasing brightly colored pills, the empty bottles of scotch, the half-full cans of paint and muddy turpentine, the forsaken toys, the cigarette butts. You have to sort and weigh and measure the volume of all the organic matter, the discards from thousands of plates: the noodles and the Cheerios and the tortillas; the pieces of pet food that have made their own gravy; the hardened jelly doughnuts, bleeding from their side wounds; the half-eaten bananas, mostly still within their peels, black and incomparably sweet in the embrace of final decay. You have to confront sticky green mountains of yard waste, and slippery brown hills of potato peels, and brittle ossuaries of chicken bones and T-bones. And then, finally, there are the "fines," the vast connecting mixture of tiny bits of paper, metal, glass, plastic, dirt, grit, and former nutrients that suffuses every landfill like a kind of grainy

* A note on terminology. Several words for the things we throw away—"garbage," "trash," "refuse," "rubbish"—are used synonymously in casual speech but in fact have different meanings. *Trash* refers specifically to discards that are at least theoretically "dry"—newspapers, boxes, cans, and so on. *Garbage* refers technically to "wet" discards—food remains, yard waste, and offal. *Refuse* is an inclusive term for both the wet discards and the dry. *Rubbish* is even more inclusive: It refers to all refuse plus construction and demolition debris. The distinction between wet and dry garbage was important in the days when cities slopped garbage to pigs, and needed to have the wet material separated from the dry; it eventually became irrevelant, but may see a revival if the idea of composting food and yard waste catches on. We will frequently use "garbage" in this book to refer to the totality of human discards because it is the word used most naturally in ordinary speech. The word is etymologically obscure, though it probably derives from Anglo-French, and its earliest associations have to do with working in the kitchen.

lymph. To understand garbage you need thick gloves and a mask and some booster shots. But the yield in knowledge—about people and their behavior as well as about garbage itself—offsets the grim working conditions.

To an archaeologist, ancient garbage pits or garbage mounds, which can usually be located within a short distance from any ruin, are always among the happiest of finds, for they contain in concentrated form the artifacts and comestibles and remnants of behavior of the people who used them. While every archaeologist dreams of discovering spectacular objects, the bread-and-butter work of archaeology involves the most common and routine kinds of discards. It is not entirely fanciful to define archaeology as the discipline that tries to understand old garbage, and to learn from that garbage something about ancient societies and ancient behaviors. The eminent archaeologist Emil Haury once wrote of the aboriginal garbage heaps of the American Southwest: "Whichever way one views the mounds—as garbage piles to avoid, or as symbols of a way of life—they nevertheless are features more productive of information than any others." When the British archaeologist Sir Leonard Woolley, in 1916, first climbed to the top of the ancient city of Carchemish, on the Euphrates River near the modern-day Turkish-Syrian border, he moistened his index finger and held it in the air. Satisfied, he scanned the region due south of the city—that is, downwind—pausing to draw on his map the location of any mounds he saw. A trench dug through the largest of these mounds revealed it to be the garbage dump Woolley was certain it was, and the exposed strata helped establish the chronological sequence for the Carchemish site as a whole. Archaeologists have been picking through ancient garbage ever since archaeology became a profession, more than a century ago, and they will no doubt go on doing so as long as garbage is produced.

Several basic points about garbage need to be emphasized at the outset. First, the creation of garbage is an unequivocal sign of a human presence. From Styrofoam cups along a roadway and urine bags on the moon there is an uninterrupted chain of garbage that reaches back more than two million years to the first "waste flake" knocked off in the knapping of the first stone tool. That the distant past often seems misty and dim is precisely because our earliest

their lives and civilizations. Many of these are little more than self-aggrandizing advertisements. The remains of the tombs, temples, and palaces of the elite are filled with personal histories as recorded by admiring relatives and fawning retainers. More such information is carved into obelisks and stelae, gouged into clay tablets, painted or printed on papyrus and paper. Historians are understandably drawn to written evidence of this kind, but garbage has often served as a kind of tattle-tale, setting the record straight.

It had long been known, for example, that French as well as Spanish forts had been erected along the coast of South Carolina during the sixteenth century, and various mounds and depressions have survived into our own time to testify to their whereabouts. Ever since the mid-nineteenth century a site on the tip of Parris Island, South Carolina, has been familiarly known as the site of a French outpost, built in 1562, that is spelled variously in old documents as Charlesfort, Charlesforte, and Charles Forte. In 1925, the Huguenot Society of South Carolina successfully lobbied Congress to erect a monument commemorating the building of Charlesfort. Subsequently, people in nearby Beaufort took up the Charlesfort theme, giving French names to streets, restaurants, and housing developments. Gift shops sold kitschy touristiana with a distinctly Gallic flavor. Those restaurants and gift shops found themselves in an awkward position when, in 1957, as a result of an analysis of discarded matter discovered at Charlesfort, a National Park Service historian, Albert Manucy, suggested that the site was of Spanish origin. Excavations begun in 1979 by the archaeologist Stanley South, which turned up such items as discarded Spanish olive jars and broken majolica pottery from Seville, confirmed Manucy's view: "Charlesfort," South established, was actually Fort San Marcos, a Spanish installation built in 1577 to protect a Spanish town named Santa Elena. (Both the fort and the town had been abandoned after only a few years.)

Garbage, then, represents physical fact, not mythology. It underscores a point that can not be too greatly emphasized: Our private worlds consist essentially of two realities—mental reality, which encompasses beliefs, attitudes, and ideas, and material reality, which is the picture embodied in the physical record. The study of garbage reminds us that it is a rare person in whom mental and material

ancestors left so little garbage behind. An appreciation of tl
plishments of the first hominids became possible only :
began making stone tools, the debris from the production
along with the discarded tools themselves, are now probec
secrets with electron microscopes and displayed in museu
garbage but as "artifacts." These artifacts serve as markers-
ingly frequent and informative markers—of how our
coped with the evolving physical and social world. Hum
are mere place-holders in time, like zeros in a long num
garbage seems to have more staying power, and a power
across the millennia that complements (and often substi
that of the written word. The profligate habits of our owi
and our own time—the sheer volume of the garbage that
and must dispose of—will make our society an open b
question is: Would we ourselves recognize our story w
told, or will our garbage tell tales about us that we as y
suspect?

That brings up a second matter: If our garbage, in the ey
future, is destined to hold a key to the past, then surely i
holds a key to the present. This may be an obvious point,
one whose implications were not pursued by scholars until
recently. Each of us throws away dozens of items every da
these items are relics of specific human activities—relics no
in their inherent nature from many of those that traditiona
ologists work with (though they are, to be sure, a bit fresher
as a whole the garbage of the United States, from its 93
households and 1.5 million retail outlets and from all of its
hospitals, government offices, and other public facilities, is
of American society. Of course, the problem with the mirror
offers is that, when encountered in a garbage can, dump, or
it is a broken one: our civilization is reflected in billions of fr
that may reveal little in and of themselves. Fitting some of th
back together requires painstaking effort—effort that a sm
ber of archaeologists and natural scientists have only just b
apply.

A third point about garbage is that it is not an assertio
physical fact—and thus may sometimes serve as a useful co
Human beings have over the centuries left many accounts de

realities completely coincide. Indeed, for the most part, the pair exist in a state of tension, if not open conflict.

Americans have always wondered, sometimes with buoyant playfulness, what their countrymen in the far future will make of Americans "now." In 1952, in a monograph he first circulated privately among colleagues and eventually published in *The Journal of Irreproducible Results,* the eminent anthropologist and linguist Joseph H. Greenberg—the man who would one day sort the roughly one thousand known Native American languages into three broad language families—imagined the unearthing of the so-called "violence texts" during an excavation of the Brooklyn Dodgers' Ebbets Field in the year A.D. 2026; what interpretation, he wondered, would be given to such newspaper reports as "Yanks Slaughter Indians" and "Reese made a sacrifice in the infield"? In 1979 the artist and writer David Macaulay published *Motel of the Mysteries,* an archaeological site-report setting forth the conclusions reached by a team of excavators in the year A.D. 4022 who have unearthed a motel dating back to 1985 (the year, Macaulay wrote, in which "an accidental reduction in postal rates on a substance called third- and fourth-class mail literally buried the North Americans under tons of brochures, fliers, and small containers called FREE"). Included in the report are illustrations of an archaeologist modeling a toilet seat, toothbrushes, and a drain stopper (or, as Macaulay describes them, "the Sacred Collar . . . the magnificent 'plasticus' ear ornaments, and the exquisite silver chain and pendant"), all assumed to be items of ritual or personal regalia. In 1982 an exhibit was mounted in New York City called "Splendors of the Sohites"—a vast display of artifacts, including "funerary vessels" (faded, dusky soda bottles) and "hermaphrodite amulets" (discarded pop-top rings), found in the SoHo section of Manhattan and dating from the Archaic Period (A.D. 1950–1961), the Classical Period (1962–1975), and the Decadent Period (1976–c.1980).

Greenberg, Macaulay, and the organizers of the Sohites exhibition all meant to have some fun, but there is an uneasy undercurrent to their work, and it is embodied in the question: What are we to make of ourselves? The Garbage Project, conceived in 1971, and officially established at the University of Arizona in 1973, was an attempt to come up with a new way of providing serious answers. It aimed to apply *real* archaeology to this very question; to see if it would be possible to investigate human behavior "from the back end," as it were. This scholarly endeavor has come to be known as garbology, and practitioners of garbology are known as garbologists. The printed citation (dated 1975) in the *Oxford English Dictionary* for the meaning of "garbology" as used here associates the term with the Garbage Project.

In the years since its founding the Garbage Project's staff members have processed more than 250,000 pounds of garbage, some of it from landfills but most of it fresh out of garbage cans in selected neighborhoods (see Figure 1-B). All of this garbage has been sorted, coded, and catalogued—every piece, from bottles of furniture polish and egg-shaped pantyhose packaging to worn and shredded clothing, crumpled bubble-gum wrappers, and the full range of kitchen waste. A unique database has been built up from these cast-offs, covering virtually every aspect of American life: drinking habits, attitudes toward red meat, trends in the use of convenience foods, the strange ways in which consumers respond to shortages, the use of contraceptives, and hundreds of other matters.*

The antecedents of the Garbage Project in the world of scholarship and elsewhere are few but various. Some are undeniably dubious. The examination of fresh refuse is, of course, as old as the human species—just watch anyone who happens upon an old campsite, or a neighbor scavenging at a dump for spare parts or furniture. The

* A question that always comes up is: What about garbage disposers? Garbage disposers are obviously capable of skewing the data in certain garbage categories, and Garbage Project researchers can employ a variety of techniques to compensate for the bias that garbage disposers introduce. Studies were conducted at the very outset of the Garbage Project to determine the discard differential between households with and without disposers, and one eventual result was a set of correction factors for various kinds of garbage (primarily food), broken down by subtype. As a general rule of thumb, households with disposers end up discarding in their trash about half the amount of food waste and food debris as households without disposers. It should be noted, however, that the fact that disposers have ground up some portion of a household's garbage often has little relevance to the larger issues the Garbage Project is trying to address. It means, for example, not that the Garbage Project's findings about the extent of food waste (see chapter three) are invalid, but merely that its estimates are conservative.

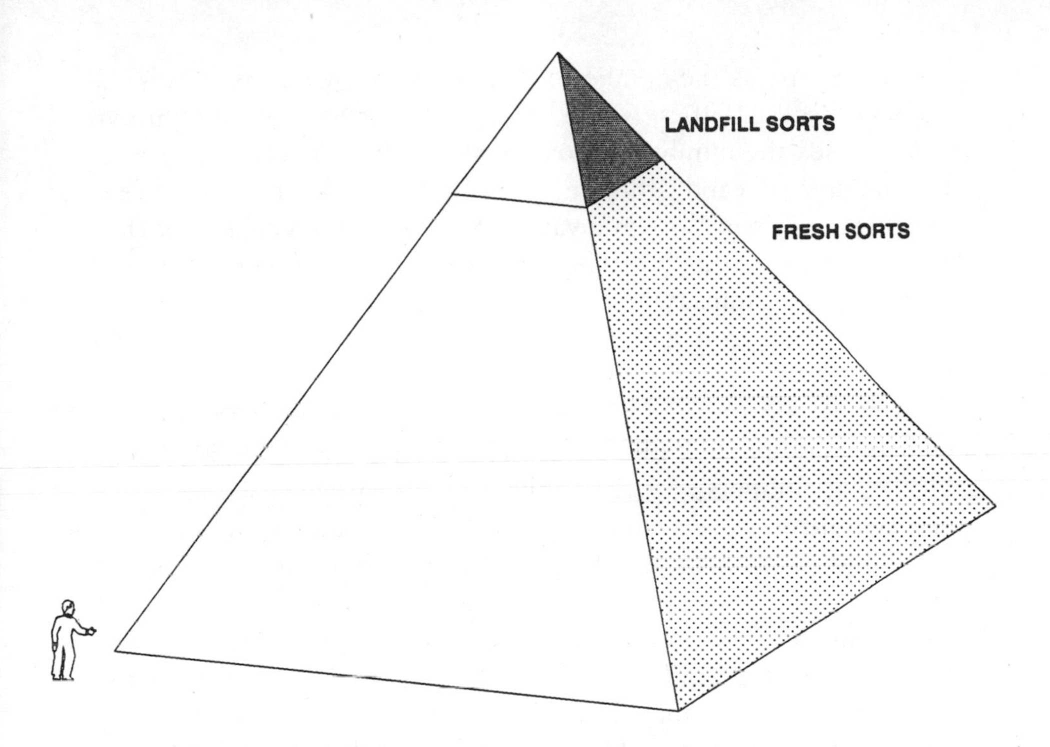

Figure **1-B.** As of mid-1991, the volume of garbage that had been sorted by Garbage Project researchers was equivalent to 1,766 cubic yards—enough to create a pyramid 56 feet square and 45 feet high. The smaller pyramid at the pinnacle shows the percent of the total sorted garbage that had been obtained from landfills, as opposed to garbage fresh from the truck.

SOURCE: Douglas Wilson, The Garbage Project

first systematic study of the components of America's garbage dates to the early 1900s and the work of the civil engineers Rudolph Hering (in New York) and Samuel A. Greeley (in Chicago), who by 1921 had gathered enough information from enough cities to compile *Collection and Disposal of Municipal Refuse,* the first textbook on urban trash management. In academe, not much happened after that for quite some time. Out in the field, however, civil engineers and solid-waste managers did now and again sort and weigh fresh garbage as it stood in transit between its source and destination, but their categories were usually simple: paper, glass, metal. No one sorted garbage into detailed categories relating to particular consumer discard patterns. No one, for example, kept track of phenomena as specific as the number of beer cans thrown away versus the number of beer bottles, or the number of orange-juice cans thrown away versus the number of pounds of freshly squeezed oranges, or the amount of candy thrown away in the week after Halloween versus the amount thrown away in the week after Valentine's Day. And no one ever dug into the final resting places of most of America's garbage: dumps (where garbage is left in the open) and sanitary landfills (where fresh garbage is covered every night with six to eight inches of soil).

Even as America's city managers over the years oversaw—and sometimes desperately attempted to cope with—the disposal of ever-increasing amounts of garbage, the study of garbage itself took several odd detours—one into the world of the military, another into the world of celebrity-watching, and a third into the world of law enforcement.

The military's foray into garbology occurred in 1941, when two enlisted men, Horace Schwerin and Phalen Golden, were forced to discontinue a survey they were conducting among new recruits about which aspects of Army life the recruits most disliked. (Conducting polls of military personnel was, they had learned, against regulations.) Schwerin and Golden had already discovered, however, that the low quality of the food was the most frequently heard complaint, and they resolved to look into this one matter with an investigation that could not be considered a poll. What Schwerin and Golden did was to station observers in mess halls to record the types of food that were most commonly wasted and the volume of waste by type of food. The result, after 2.4 million man-meals had been observed, was a textbook example of how garbage studies can produce not only behavioral insights but also practical benefits. Schwerin and Golden discovered that 20 percent of the food prepared for Army mess halls was eventually thrown away, and that one reason for this was simply excess preparation. Here are some more of their findings, as summarized in a wartime article that appeared in the *The Saturday Evening Post:*

> Soldiers ate more if they were allowed to smoke in the mess hall. They ate more if they went promptly to table instead of waiting on line outside—perhaps because the food became cold. They

ate more if they fell to on their own initiative instead of by command. They cared little for soups, and 65 percent of the kale and nearly as much of the spinach went into the garbage can. Favorite desserts were cakes and cookies, canned fruit, fruit salad, and gelatin. They ate ice cream in almost any amount that was served to them.

"That, sergeant, is an excellent piece of work," General George C. Marshall, the Army chief of staff, told Horace Schwerin after hearing a report by Schwerin on the research findings. The Army adopted many of Schwerin and Golden's recommendations, and began saving some 2.5 million pounds of food a day. It is perhaps not surprising to learn that until joining the Army Horace Schwerin had been in market research, and, among other things, had helped CBS to perfect a device for measuring audience reaction to radio shows.

The origins of an ephemeral branch of garbage studies focused on celebrities—"peeping-Tom" garbology, one might call it—seem to lie in the work of A. J. Weberman. Weberman was a gonzo journalist and yippie whose interest in the songs of Bob Dylan, and obsession with their interpretation, in 1970 prompted him to begin stealing the garbage from the cans left out in front of Dylan's Greenwich Village brownstone on MacDougal Street. Weberman didn't find much—some soiled Pampers, some old newspapers, some fast-food packaging from a nearby Blimpie Base, a shopping list with the word vanilla spelled "vannilla." He did, however, stumble into a brief but highly publicized career. This self-proclaimed "garbage guerrilla" quickly moved on to Neil Simon's garbage (it included a half-eaten bagel, scraps of lox, the Sunday *Times)*, Muhammad Ali's (an empty can of Luck's collard greens, an empty roach bomb), and Abbie Hoffman's (a summons for hitchhiking, an unused can of deodorant, an estimate of the cost for the printing of *Steal This Book*, and the telephone numbers of Jack Anderson and Kate Millet). Weberman revealed many of his findings in an article in *Esquire* in 1971. It was antics such as his that inspired a prior meaning of the term "garbology," one very different from the definition established today.

Weberman's work inspired other garbage guerrillas. In January of 1975, the *Detroit Free Press* Sunday magazine reported on the findings from its raids on the garbage of several city notables, including the mayor, the head of the city council, the leader of a right-wing group, a food columnist, a disk jockey, and a prominent psychiatrist. Nothing much was discovered that might be deemed out of the ordinary, save for some of the contents of the garbage taken from a local Hare Krishna temple: a price tag from an Oleg Cassini garment, for example, and four ticket stubs from the Bel-Aire Drive-In Theater, which at the time was showing *Horrible House on the Hill* and *The Night God Screamed.* Six months after the *Free Press* exposé, a reporter for the *National Enquirer,* Jay Gourley, drove up to 3018 Dumbarton Avenue, N.W., in Washington, D.C., and threw the five garbage bags in front of Secretary of State Henry A. Kissinger's house into the trunk of his car. Secret Service agents swiftly blocked Gourley's departure, but after a day of questioning allowed him to proceed, the garbage still in the trunk. Among Gourley's finds: a crumpled piece of paper with a dog's teeth marks on it, upon which was written the work schedules of the Secret Service agents assigned to guard the Secretary; empty bottles of Seconal and Maalox; and a shopping list, calling for a case of Jack Daniel's, a case of Ezra Brooks bourbon, and a case of Cabin Still bourbon. Gourley later returned most of the garbage to the Kissingers—minus, he told reporters, "several dozen interesting things."

After the Kissinger episode curiosity about the garbage of celebrities seems to have abated. In 1977 the *National Enquirer* sent a reporter to poke through the garbage of President Jimmy Carter's press secretary, Jody Powell. The reporter found so little of interest that the tabloid decided not to publish a story. In 1980 Secret Service agents apprehended A. J. Weberman as he attempted to abduct former President Richard Nixon's garbage from behind an apartment building in Manhattan. Weberman was released, without the garbage.

The third detour taken by garbage studies involves police work. Over the years, law enforcement agents looking for evidence in criminal cases have also been more-than-occasional students of garbage; the Federal Bureau of Investigation in particular has spent considerable time poring over the household trash of people in whom it maintains a professional interest. ("We take it on a case-by-case basis," an FBI spokesman says.) One of the biggest criminal cases involving garbage began in 1975 and involved Joseph "Joe Bananas"

Bonanno, Sr., a resident of Tucson at the time and a man with alleged ties to organized crime that were believed to date back to the days of Al Capone. For a period of three years officers of the Arizona Drug Control District collected Bonanno's trash just before the regular pickup, replacing it with "fake" Bonanno garbage. (Local garbagemen were not employed in the operation because some of them had received anonymous threats after assisting law enforcement agencies in an earlier venture.) The haul in evidence was beyond anyone's expectations: Bonanno had apparently kept detailed records of his various transactions, mostly in Sicilian. Although Bonanno had torn up each sheet of paper into tiny pieces, forensic specialists with the Drug Control District, like archaeologists reconstructing ceramic bowls from potsherds, managed to reassemble many of the documents and with the help of the FBI got them translated. In 1980 Bonanno was found guilty of having interfered with a federal grand jury investigation into the business operations of his two sons and a nephew. He was eventually sent to jail.

Unlike law-enforcement officers or garbage guerrillas, the archaeologists of the Garbage Project are not interested in the contents of any particular individual's garbage can. Indeed, it is almost always the case that a given person's garbage is at once largely anonymous and unimaginably humdrum. Garbage most usefully comes alive when it can be viewed in the context of broad patterns, for it is mainly in patterns that the links between artifacts and behaviors can be discerned.

The seed from which the Garbage Project grew was an anthropology class conducted at the University of Arizona in 1971 that was designed to teach principles of archaeological methodology. The University of Arizona has long occupied a venerable place in the annals of American archaeology and, not surprisingly, the pursuit of archaeology there to this day is carried on in serious and innovative ways. The class in question was one in which students undertook independent projects aimed precisely at showing links between various kinds of artifacts and various kinds of behavior. For example, one student, Sharon Thomas, decided to look into the relationship between a familiar motor function ("the diffusion pattern of ketchup over hamburgers") and a person's appearance, as manifested in clothing. Thomas took up a position at "seven different hamburger dispensaries" and, as people came in to eat, labeled them "neat" or "sloppy" according to a set of criteria relating to the way they dressed. Then she recorded how each of the fifty-seven patrons she studied—the ones who ordered hamburgers—poured ketchup over their food. She discovered that sloppy people were far more likely than neat people to put ketchup on in blobs, sometimes even stirring it with their fingers. Neat people, in contrast, tended to apply the ketchup in patterns: circles, spirals, and crisscrosses. One person (a young male neatly dressed in a body shirt, flared pants, and patent-leather Oxfords) wrote with ketchup what appeared to be initials.

Two of the student investigations, conducted independently by Frank Ariza and Kelly Allen, led directly to the Garbage Project. Ariza and Allen, wanting to explore the divergence between (or correlation of) mental stereotypes and physical realities, collected garbage from two households in an affluent part of Tucson and compared it to garbage from two households in a poor and, as it happens, Mexican-American part of town. The rich and poor families, each student found, ate about the same amount of steak and hamburger, and drank about the same amount of milk. But the poor families, they learned, bought more expensive child-education items. They also bought more household cleansers. What did such findings mean? Obviously the sample—involving only four households in all—was too small for the results even to be acknowledged as representative, let alone to provide hints as to what lay behind them. However, the general nature of the research effort itself—comparing garbage samples in order to gauge behavior (and, what is more, gauging behavior unobtrusively, thereby avoiding one of the great biases inherent in much social science)—seemed to hold great promise.

A year later, in 1972, university students, under professorial direction, began borrowing samples of household garbage from different areas of Tucson, and sorting it in a lot behind a dormitory. The Garbage Project was under way. In 1973, the Garbage Project entered into an arrangement with the City of Tucson, whereby the Sanitation Division, four days a week, delivered five to eight randomly selected household pickups from designated census tracts to an analysis site that the Division set aside for the Project's sorters at a maintenance yard. (Wilson Hughes, who as mentioned earlier is the Garbage Project's co-director, was one of the first undergraduate garbage sorters.) In 1984 operations were moved to an enclosure where many of the university's dumpsters are parked, across the street from Arizona Stadium.

The excavation of landfills would come much later in the Garbage Project's history, when to its focus on issues of garbage and human behavior it added a focus on issues of garbage management. The advantage in the initial years of sorting fresh garbage over excavating landfills was a basic but important one: In landfills it is often quite difficult and in many cases impossible to get some idea, demographically speaking, of the kind of neighborhood from which any particular piece of garbage has come. The value of landfill studies is therefore limited to advancing our understanding of garbage in the aggregate. With fresh garbage, on the other hand, one can have demographic precision down to the level of a few city blocks, by directing pickups to specific census districts and cross-tabulating the findings with census data.

Needless to say, deciding just which characteristics of the collected garbage to pay attention to posed a conceptual challenge, one that was met by Wilson Hughes, who devised the "protocol" that is used by the Garbage Project to this day. Items found in garbage are sorted into one of 150 specific coded categories (see Figure 1-C) that can in turn be clustered into larger categories representing food (fresh food versus prepared, health food versus junk food), drugs, personal and household sanitation products, amusement-related or educational materials, communications-related materials, pet-related materials, yard-related materials, and hazardous materials. For each item the following information is recorded on a standardized form: the date on which it was collected; the census tract from which it came; the item code (for example, 001, which would be the code for "Beef"); the item's type (for example, "chuck"); its original weight or volume (in this case, derived from the packaging); its cost (also from the packaging); material composition of container; brand (if applicable); and the weight of any discarded food (if applicable). The information garnered over the years from many thousands of such forms, filled out in pursuit of a wide variety of research objectives, consti-

GARBAGE ITEM CODE LIST

The Garbage Project
University of Arizona

Item	Code
BEEF *	001
OTHER MEAT (not bacon) *	002
CHICKEN	003
OTHER POULTRY	004
FISH (fresh, frozen, canned, dried) *	005
CRUSTACEANS & MOLLUSKS (shrimp, clams, etc.)	006
T.V.P. TYPE FOODS *	007
UNKNOWN MEAT	008
CHEESE (including cottage cheese)	010
MILK *	011
ICE CREAM (also ice milk, sherbet) *	012
OTHER DAIRY (not butter)	013
EGGS (regular, powdered, liquid) *	014
BEANS (not green beans) *	015
NUTS	016
PEANUT BUTTER	017
FATS: Saturated *	018
Unsaturated *	019
Bacon, saltpork *	020
Meat trimming	021
CORN (also corn meal and masa) *	022
FLOUR (also pancake mix) *	023
RICE *	024
OTHER GRAIN (barley, wheat germ, etc.)	025
NOODLES (pasta)	026
WHITE BREAD	027
DARK BREAD	028
TORTILLAS *	029
DRY CEREALS: Regular	030
High Sugar (first ingredient only)	031
COOKED CEREALS (instant or regular)	032
CRACKERS	033
CHIPS (also pretzels)	034
UNKNOWN PRODUCE *	040
FRESH VEGETABLES *	041
CANNED VEGETABLES (dehydrated also) *	042
FROZEN VEGETABLES *	043
POTATO PEEL *	044
FRESH FRUIT *	045
CANNED FRUIT (dehydrated also) *	046
FROZEN FRUIT *	047
FRUIT PEEL *	048
RELISH, PICKLES, OLIVES *	049
SYRUP, HONEY, JELLIES, MOLASSES	051
PASTRIES (cookies, cakes and mix, pies, etc.) *	052
SUGAR *	053
ARTIFICIAL SWEETENERS	054
CANDY *	055
SALT *	056
SPICES & FLAVORINGS (catsup, mustard, pepper, etc.) *	057
BAKING ADDITIVES (yeast, baking powder, etc.)	058
POPSICLES	060
PUDDING	061
GELATIN	062
INSTANT BREAKFAST	063
DIPS (for chips)	064
NON-DAIRY CREAMERS & WHIPS	065
HEALTH FOODS *	066
SLOPS *	069
REGULAR COFFEE (instant or ground) *	070
DECAF COFFEE	071
EXOTIC COFFEE *	072
TEA *	073
CHOCOLATE DRINK MIX OR TOPPING	074
FRUIT OR VEG JUICE (canned or bottled)	075
FRUIT JUICE CONCENTRATE	076
FRUIT DRINK, pdr or lqud (Tang, Koolaid, Hi-C) *	077
DIET SODA	078
REGULAR SODA	079
COCKTAIL MIX (carbonated)	080
COCKTAIL MIX (non-carb. liquid)	081
COCKTAIL MIX (powdered)	082
PREMIXED COCKTAILS (alcoholic)	083
SPIRITS (booze)	084
WINE (still & sparkling)	085
BEER *	086
BABY FOOD & JUICE *	087
BABY CEREAL (pablum)	088
BABY FORMULA (liquid) *	089
BABY FORMULA (powdered) *	090
PET FOOD (dry)	091
PET FOOD (canned or moist)	092
TV DINNERS (also pot pies)	094
TAKE OUT MEALS	095
SOUPS *	096
GRAVY & SPECIALTY SAUCES *	097
PREPARED MEALS (canned or packaged) *	098
VITAMIN PILLS AND SUPPLEMENTS (commercial) *	100
PRESCRIBED DRUGS (prescribed vitamins)	101
ASPIRIN *	102
COMMERCIAL STIMULANTS AND DEPRESSANTS *	103
COMMERCIAL REMEDIES *	104
ILLICIT DRUGS *	105
COMMERCIAL DRUG PARAPHENALIA	106
ILLICIT DRUG PARAPHENALIA	107
CONTRACEPTIVES: MALE	108
FEMALE	109
BABY SUPPLIES (diapers, etc.) *	111
INJURY ORIENTED (iodine, bandaids, etc.)	112
PERSONAL SANITATION *	113
COSMETICS *	114
CIGARETTES (butts)	123
CIGARETTES (pack) *	124
CIGARETTES (carton) *	125
CIGARS	126
PIPE, CHEWING TOBACCO, LOOSE TOBACCO	127
ROLLING PAPERS (also smoking items)	128
HOUSEHOLD & LAUNDRY CLEANERS *	131
HOUSEHOLD CLEANING TOOLS (not detergents)	132
HOUSEHOLD MAINT. ITEMS (paint, wood, etc.)	133
COOKING & SERVING AIDS	134
TISSUE CONTAINER	135
TOILET PAPER CONTAINER	136
NAPKIN CONTAINER	137
PAPER TOWEL CONTAINER	138
PLASTIC WRAP CONTAINER	139
BAGS (paper or plastic) *	140
BAG CONTAINER	141
ALUMINUM FOIL SHEETS	142
ALUMINUM FOIL PACKAGE	143
WAX PAPER PACKAGE	144
MECHANICAL APPLIANCE (tools)	147
ELECTRICAL APPLIANCE AND ITEMS	148
AUTO SUPPLIES	149
FURNITURE	150
CLOTHING: CHILD *	151
ADULT *	152
CLOTHING CARE ITEMS (shoe polish, thread)	153
DRY CLEANING (laundry also)	154
PET MAINTENANCE (litter)	155
PET TOYS	156
GATE RECEIPTS (tickets)	157
HOBBY RELATED ITEMS	158
PHOTO SUPPLIES	159
HOLIDAY VALUE (non-food) *	160
DECORATIONS (non holiday)	161
PLANT AND YARD MAINT	162
STATIONERY SUPPLIES	163
JEWELRY	164
CHILD SCHOOL RELATED PAPERS *	171
CHILD EDUC. BOOKS (non-fiction)	172
CHILD EDUC. GAMES (toys)	173
CHILD AMUSEMENT READING	174
CHILD AMUSEMENT TOYS (games)	175
ADULT BOOKS (non-fiction)	176
ADULT BOOKS (fiction)	177
ADULT AMUSEMENT GAMES	178
LOCAL NEWSPAPERS *	181
NEWSPAPERS (other city, national) *	182
ORGANIZATIONAL NEWSPAPERS OR MAGAZINES (also religion) *	183
GENERAL INTEREST MAGAZINES *	184
SPECIAL INTEREST MAGAZINE OR NEWSPAPER *	185
ENTERTAINMENT GUIDE (TV Guide, etc.)	186
MISCELLANEOUS ITEMS (specify on back of sheet) *	190

* See Special Notes

Figure 1-C. Garbage Project sorters use the codes displayed here to begin the process of transforming raw garbage into data. The code numbers are supplemented on the recording forms by much more detailed information, such as a discarded item's brand name (if applicable), its type, its weight, the census tract from which it originated, the date of collection, and so on. Not shown here are several pages of specialized instructions, such as this for the item with code number 044: "Do *not* count individual peels; weigh them as a group."

SOURCE: The Garbage Project

tutes the Garbage Project's database. It has all been computerized and amounts to some two million lines of data drawn from some fifteen thousand household-refuse samples. The aim here has been not only to approach garbage with specific questions to answer or hypotheses to prove but also to amass sufficient quantities of information, in a systematic and open-minded way, so that with the data on hand Garbage Project researchers would be able to answer any future questions or evaluate any future hypotheses that might arise. In 1972 garbage was, after all, still terra incognita, and the first job to be done was akin to that undertaken by the explorers Lewis and Clark.

From the outset the Garbage Project has had to confront the legal and ethical issues its research involves: Was collecting and sorting someone's household garbage an unjustifiable invasion of privacy? This very question has over the years been argued repeatedly in the courts. The Fourth Amendment unequivocally guarantees Americans protection from unreasonable search and seizure. Joseph Bonanno, Sr., tried to invoke the Fourth Amendment to prevent his garbage from being used as evidence. But garbage placed in a garbage can in a public thoroughfare, where it awaits removal by impersonal refuse collectors, and where it may be picked over by scavengers looking for aluminum cans, by curious children or neighbors, and by the refuse collectors themselves (some of whom do a thriving trade in old appliances, large and small), is usually considered by the courts to have been abandoned. Therefore, the examination of the garbage by outside parties cannot be a violation of a constitutional right. In the Bonanno case, U.S. District Court Judge William Ingram ruled that investigating garbage for evidence of a crime may carry a "stench," but was not illegal. In 1988, in *California* v. *Greenwood*, the U.S. Supreme Court ruled by a margin of six to two that the police were entitled to conduct a warrantless search of a suspected drug dealer's garbage—a search that led to drug paraphenalia, which led in turn to warrants, arrests, and convictions. As Justice Byron White has written, "The police cannot reasonably be expected to avert their eyes from evidence of criminal activity that could have been observed by any member of the public."

Legal issues aside, the Garbage Project has taken pains to ensure that those whose garbage comes under scrutiny remain anonymous.

Before obtaining garbage for study, the Project provides guarantees to communities and their garbage collectors that nothing of a personal nature will be examined and that no names or addresses or other personal information will be recorded. The Project also stipulates that all of the garbage collected (except aluminum cans, which are recycled) will be returned to the community for normal disposal.

As noted, the Garbage Project has now been sorting and evaluating garbage, with scientific rigor, for two decades. The Project has proved durable because its findings have supplied a fresh perspective on what we know—and what we think we know—about certain aspects of our lives. Medical researchers, for example, have long made it their business to question people about their eating habits in order to uncover relationships between patterns of diet and patterns of disease. These researchers have also long suspected that people—honest, well-meaning people—may often be providing information about quantities and types and even brands of food and drink consumed that is not entirely accurate. People can't readily say whether they trimmed 3.3 ounces or 5.4 ounces of fat off the last steak they ate, and they probably don't remember whether they had four, five, or seven beers in the previous week, or two eggs or three. The average person just isn't paying attention. Are there certain patterns in the way in which people wrongly "self-report" their dietary habits? Yes, there are, and Garbage Project studies have identified many of them.

Garbage archaeologists also know how much edible food is thrown away; what percentage of newspapers, cans, bottles, and other items aren't recycled; how loyal we are to brand-name products and which have earned the greatest loyalty; and how much household hazardous waste is carted off to landfills and incinerators. From several truckloads of garbage and a few pieces of ancillary data—most importantly, the length of time over which the garbage was collected—the Garbage Project staff can reconstruct the community from which it came with a degree of accuracy that the Census Bureau might in some neighborhoods be unable to match.

Garbage also exposes the routine perversity of human ways. Garbage archaeologists have learned, for example, that the volume of

garbage that Americans produce expands to fill the number of receptacles that are available to put it in. They have learned that we waste more of what is in short supply than of what is plentiful; that attempts by individuals to restrict consumption of certain foodstuffs are often counterbalanced by extra and inadvertent consumption of those same foodstuffs in hidden form; and that while a person's memory of what he has eaten and drunk in a given week is inevitably wide of the mark, his guess as to what a family member or even neighbor has eaten and drunk usually turns out to be more perceptive.

Some of the Garbage Project's research has prompted unusual forays into arcane aspects of popular culture. Consider the matter of those "amulets" worn by the Sohites—that is, the once-familiar detachable pop-top pull tab. Pull tabs first became important to the Garbage Project during a study of household recycling practices, conducted on behalf of the federal Environmental Protection Agency during the mid-1970s. The question arose: If a bag of household garbage contained no aluminum cans, did that mean that the household didn't dispose of any cans or that it had recycled its cans? Finding a way to answer that question was essential if a neighborhood's recycling rate was to be accurately determined. Pull tabs turned out to hold the key. A quick study revealed that most people did not drop pull tabs into the cans from which they had been wrenched; rather, the vast majority of people threw the tabs into the trash. If empty cans were stored separately for recycling, the pull tabs still went out to the curb with the rest of the garbage. A garbage sample that contained several pull tabs but no aluminum cans was a good bet to have come from a household that recycled.

All this counting of pull tabs prompted a surprising discovery one day by a student: Pull tabs were not all alike. Their configuration and even color depended on what kind of beverage they were associated with and where the beverage had been canned. Armed with this knowledge, Garbage Project researchers constructed an elaborate typology of pull tabs, enabling investigators to tease out data about beverage consumption—say, beer versus soda, Michelob versus Schlitz—even from samples of garbage that contained not a single can (see Figure 1-D). Detachable pull tabs are no longer widely used in beverage cans, but the pull-tab typology remains useful even

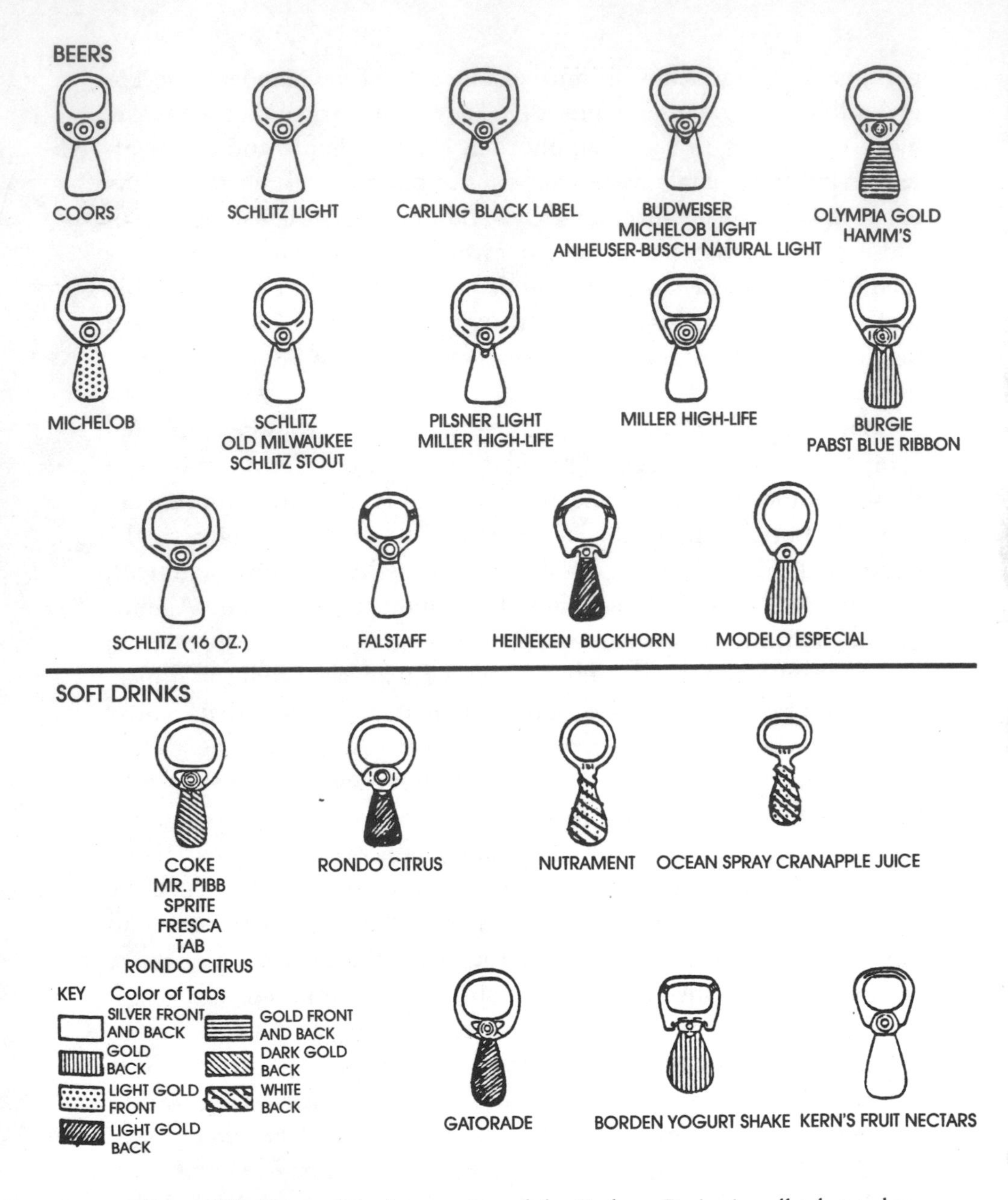

Figure 1-D. Shown here is a portion of the Garbage Project's pull-tab typology for the city of Tucson, underscoring the widespread stylistic variation even in relatively standardized everyday items. The typology was originally developed to assist in a study of recycling behavior for the Environmental Protection Agency.

SOURCE: The Garbage Project

now. Among other things, in the absence of such evidence of chronology as a newspaper's dateline, pull tabs can reliably help to fix the dates of strata in a landfill. In archaeological parlance objects like these that have been widely diffused over a short period of time, and then abruptly disappear, are known as horizon markers.

The unique "punch-top" on Coors beer cans, for example, was used only between March of 1974 and June of 1977. (It was abandoned because some customers complained that they cut their thumbs pushing the holes open.) In landfills around the country, wherever Coors beer cans were discarded, punch-top cans not only identify strata associated with a narrow band of dates but also separate two epochs one from another. One might think of punch-tops playfully as the garbage equivalent of the famous iridium layer found in sediment toward the end of the Cretaceous Era, marking the moment (proponents of the theory believe) when a giant meteor crashed into the planet Earth, exterminating the dinosaurs.

All told, the Garbage Project has conducted nine full-scale excavations of municipal landfills in the United States and two smaller excavations associated with special projects. In the fall of 1991 it also excavated four sites in Canada, the data from which remains largely unanalyzed (and is not reflected in this book). The logistics of the landfill excavations are complex, and they have been overseen in all cases by Wilson Hughes. What is involved? Permission must be obtained from a raft of local officials and union leaders; indemnification notices must be provided to assure local authorities that the Garbage Project carries sufficient insurance against injury; local universities must be scoured for a supply of students to supplement the Garbage Project team; in many cases construction permits, of all things, must be obtained in advance of digging. There is also the whole matter of transportation, not only of personnel but also of large amounts of equipment. And there is the matter of personal accommodation and equipment storage. The time available for excavation is always limited, sometimes extremely so; the research program must be compressed to fit it, and the staff must be "tasked" accordingly. When the excavation has been completed the samples need to be packed and shipped—frequently on ice—back to headquarters or to specialized laboratories. All archaeologists will tell you that field work is mostly laborious, not glamorous; a landfill excavation is archaeology of the laborious kind.

For all the difficulties they present, the Garbage Project's landfill digs have acquired an increasing timeliness and relevance as concerns about solid-waste disposal have grown. Even as the Garbage Project has trained considerable attention on garbage as an analytical tool it has also taken up the problem of garbage itself—garbage as a problem, garbage as symbolized by *Mobro 4000,* the so-called "garbage barge," which sailed from Islip, Long Island, on March 22, 1987, and spent the next fifty-five days plying the seas in search of a place to deposit its 3,168 tons of cargo. Strange though it may seem, although more than 70 percent of America's household and commercial garbage ends up in landfills, very little reliable data existed until recently as to a landfill's contents and biological dynamics. Much of the conventional wisdom about garbage disposal consists of assertions that turn out, upon investigation, to be simplistic or misleading: among them, the assertion that, as trash, plastic, foam, and fast-food packaging are causes for great concern, that biodegradable items are always more desirable than nonbiodegradable ones, that on a per capita basis the nation's households are generating a lot more garbage than they used to, and that we're physically running out of places to put landfills.

This is not to say that garbage isn't a problem in need of serious attention. It is. But if they are to succeed, plans of action must be based on garbage realities. The most critical part of the garbage problem in America is that our notions about the creation and disposal of garbage are often riddled with myth. There are few other subjects of public significance on which popular and official opinion is so consistently misinformed.

This book is a summary of the research conducted and discoveries made by the Garbage Project over the course of two decades. In the following chapters we will first step back for a moment into human history and look at the place of garbage in it. We will then move on to some of the insights into human behavior that an examination of garbage can yield. We will next venture inside a landfill and examine its actual contents, and try to understand what happens—and doesn't happen—to the garbage that winds up there. We will conclude by discussing a few issues that receive a great deal of vocal

attention, such as incineration and recycling. Along the way there will be digressions large and small—about disposable diapers, about the demographics of garbage, about many other odd or essential things.

Gaps—large gaps—remain in our knowledge of garbage, and of how human behavior relates to it, and of how best to deal with it. But a lighted candle has at least been seized and thrust inside the antechamber.

CHAPTER 2

GARBAGE AND HISTORY

The sixty-second advertisement for AT&T known in the industry by the name "The Dig" was first broadcast on television at the primest of times: at 8:17 p.m. on October 29, 1989, during the first episode of ABC's made-for-TV movie *The Final Days,* a much-publicized and controversial docudrama recounting Richard Nixon's fall from power. "The Dig" was produced by Ayer Advertising, in New York City, and its debut was seen by some 14.6 million Americans.

Archaeologists watching *The Final Days* probably got a bit of a lump in their throats when the commercial appeared on the screen. Here is what it showed: The foundations of a skyscraper are being dug when cries of discovery suddenly bring work to a halt. A team of archaeologists rushes to the scene, led by a young woman in a yellow hard hat. There follows a stylish, quick-cut sequence: Troweling exposes the wooden skeleton of an old ship; the skeletal frame is transposed to a computer screen, which rotates it like a design on the drawing boards for a new car; finally, dirt is excitedly brushed aside to reveal the ship's figurehead in tarnished splendor. Through it all there are a lot of phone calls. "This is big. Really big," someone

says. The whole episode was exhilarating, and it was also good to see the virtues of conservation being celebrated—even if one knew that the ship's "frame" and the ship's "figurehead" were not real artifacts but rather stage props molded from Styrofoam and painted to look like wood; even, indeed, if one knew that the excavation on which the commercial was based involved a ship that did not have a figurehead.

The real excavation took place at 175 Water Street, in the financial district of Manhattan, in January of 1982, after workers, in advance of a major construction project involving the National Westminster Bank, dug several deep, four-foot by ten-foot exploratory holes at randomly selected places on the site so that archaeologists could check for any significant archaeological remains. (Archaeological testing of this kind is now mandatory in many states and municipalities.) As one of these holes was dug the mud siding sloughed away and exposed a ship's frame, ten feet below street level.

The archaeologist in charge of the excavation that ensued was Sheli Smith, now the curator of the Los Angeles Maritime Museum. For six semesters she had been one of the Garbage Project's most assiduous garbage sorters, once telling a *Wall Street Journal* reporter that she sorted garbage "to relax." In her capacity as a garbage sorter Smith appeared in 1975 on the television show "To Tell the Truth," and managed to elude discovery by all four of the panelists. Her success had everything to do with a special manicure she arranged for herself before the show: "No one with nails like that would ever sort garbage," the panelist Peggy Cass confidently stated.

The ship that Smith uncovered at 175 Water Street—now known as the Ronson ship, after Howard Ronson, the developer of the property—was once a proud, three-masted merchantman. It may have been in the tobacco trade for a time, and the presence of certain species of taredo worms in its furring suggests that the ship sailed at least once to the South Seas. But sometime around 1750, her masts gone, the ship was positioned on a tidal flat abutting what was then lower Manhattan's shoreline to become part of a retaining wall. In doing so, it also became part of the process by which Manhattan's shoreline has steadily encroached on surrounding waterways. The ship was filled with ballast and sludge and then heaped with building debris and assorted garbage, including the castoff leather and tacks of a cobbler, and the castoff cow heads and pig heads of a victualler. "From all those years of sorting garbage," Sheli Smith recalls, "I knew what we had right from the start."

To archaeologists, the Ronson ship was a find of major importance, because the vessel is the first colonial merchant ship to be discovered that they have had the opportunity to preserve and study. To garbologists, the unearthing of the ship was heralded for an entirely different reason. Its discovery reminds us that, over time, the world of garbage is characterized by continuity. The Ronson ship has obvious forebears, to give but one example, in the wharves that lined the channel connecting the Tiber River and Rome's port of Ostia—wharves made from derelict scows that had been packed with garbage and topped with concrete. And it has obvious descendants in those shoreline extensions of land, built over many years out of hard-packed garbage, that are today the sites of such places as LaGuardia Airport, in New York City, and the John F. Kennedy Library, in Boston.

The examples here may seem trivial, and yet the fact is—look where one may—that the history of garbage consists largely of a relatively few long, simple, durable strands of behavior. Our relentless if understandable present-mindedness often keeps us from seeing that our own practices with respect to garbage are, far from being somehow novel and unique, deeply rooted in the ways of our ancestors—a fact that might at least offer some modest psychological comfort, even if it makes the task of garbage disposal in our own time no easier. That same present-mindedness also blinds us to the ways in which, for better or worse, our latter-day behaviors and practices *are* unique.

Throughout most of time human beings disposed of garbage in a very convenient manner: simply by leaving it where it fell. To be sure, they sometimes tidied up their sleeping and activity areas, but that was about all. This disposal scheme functioned adequately because hunter-gatherers frequently abandoned their campgrounds to follow game or find new stands of plants (and, of course, because there weren't all that many hunter-gatherers to begin with). When modern hunter-gatherers, like the aborigines of the Australian outback, are provided with government tract housing, one of the immediate problems they face is that of garbage disposal. Accustomed to simply moving on several times a year for any one of a handful of reasons, including an unendurable accumulation of garbage within the trash perimeter of their temporary camps, some aborigines have been at a loss when encouraged by authorities to settle in a more stable sort of camp—namely, a house. As James F. O'Connell, an American anthropologist who works among the Alyawara tribe in Australia, has noted, "Where housing is permanent, the refuse rather than the people will have to be moved, which means a major readjustment in present behavior patterns."

As such habits suggest, our species faced its first garbage crisis when human beings became sedentary animals. The archaeologist Gordon R. Willey, who in the late 1940s conducted in Peru the first extensive archaeological study of regional settlement patterns over time, has argued (only partly in jest) that *Homo sapiens* may have been propelled along the path to civilization by his need for a degree of organization sufficiently sophisticated, and a class structure suitably stratified, to make possible the disposal of mounting piles of debris.

There are no ways of dealing with garbage that haven't been familiar, in essence, for thousands of years, although as the species has advanced, people have introduced refinements. The basic methods of garbage disposal are four: dumping it, burning it, turning it into something that can be useful (recycling), and minimizing the volume of material goods—future garbage—that comes into existence in the first place (this last is known technically in the garbage field as "source reduction"). Any civilization of any complexity has used all four procedures simultaneously to one degree or another.

The ancient Maya, for instance, deposited much of their organic waste in what we would today call open dumps. These dumps probably experienced the occasional explosion as a result of the methane gas building up inside them, and some of the piles of garbage would have been continually burning or smoldering, making room for more garbage to be dumped. The Maya also recycled inorganic garbage—mainly broken pottery, grinding stones, and cut stone from the façades of old buildings—by using it as fill in temples or for other building projects. And the Maya were adept at source reduction. In the Late Postclassic period—after A.D. 1200—they drastically curbed demand for richly ornamented ceramics, ritual paraphernalia, and body ornaments, and thereby achieved a significant savings in scarce or costly resources. They did so (perhaps in the face of economic decline) by the simple expedient of abandoning the practice of burying the dead with new or intact pottery, tools, and jewelry, and burying them instead with objects that were broken. In addition they substituted "fake" for original art—for example, clay beads covered with gold foil instead of beads of solid gold.

Human beings have been deploying the four main weapons against garbage for so long that they are by now well aware of each method's relative convenience. Not surprisingly, a human being's first inclination is always to dump. From prehistory through the present day, dumping has been the means of disposal favored everywhere, including within cities. Archaeological excavations of hard-packed dirt and clay floors—the most common type of ancient living surface—usually recover an amplitude of small finds, suggesting that many bits of garbage that fell on the floor were trampled into the dirt or were brushed into corners and along the edge of walls by the traffic patterns of the occupants. (This dispersal of garbage to the edges of an occupied space is known to archaeologists as the "fringe effect.") The archaeologist C. W. Blegen, who dug into Bronze Age Troy during the 1950s, found that the floors of its buildings had periodically become so littered with animal bones and small artifacts that "even the least squeamish household felt that something had to be done." This was normally accomplished, Blegen discovered,

> not by sweeping out the offensive accumulation, but by bringing in a good supply of fresh clean clay and spreading it out thickly to cover the noxious deposit. In many a house, as demonstrated by the clearly marked stratification, this process was repeated time after time until the level of the floor rose so high that it was necessary to raise the roof and rebuild the doorway.

Eventually, of course, buildings had to be demolished altogether, the old mud-brick walls knocked in to serve as the foundations of new mud-brick buildings. Over time the ancient cities of the Middle East rose high above the surrounding plains on massive mounds, called *tells,* which contained the ascending remains of centuries, even millennia, of prior occupation (see Figure 2-A). In 1973 Charles Gunnerson, a civil engineer with the U.S. Department of Commerce's Environmental Research Laboratories, calculated that the rate of elevation due to debris accumulation in Troy was about 4.7 feet per century. If the idea of a city rising above its gradually accumulating fill and debris at this rate seems extraordinary, recall the depth below street level at which Sheli Smith's ship was found. "Street level" on the island of Manhattan today is typically six to fifteen feet higher than it was when Peter Minuit lived there; in some places it is as much as thirty feet higher. Nowadays, needless to say, the fill used in construction on Manhattan is not normally garbage, but Gunnerson calculated that if all of the garbage from Manhattan that is currently sent to Fresh Kills and all the construction and demolition debris from Manhattan that is currently dumped at sea were instead spread out evenly over the island, the rate of accumulation per century would be exactly the same as that of ancient Troy (see Figure 2-B).

At Troy and elsewhere, of course, not all trash was kept indoors. The larger pieces of garbage and debris were thrown into the streets. As structures became multistoried the practice of throwing garbage from upper floors to the ground below became commonplace. Up until relatively modern times, once garbage landed in the streets, semidomesticated animals, usually pigs and dogs, ate up the food scraps. Even after the advent of modern landfills—and even in the United States—the "slopping" of garbage to pigs continued in a major way, though it was done on farms and not in the streets. A survey of 557 American cities in 1930 found that about 40 percent of them still saved their wet garbage for the purpose of slopping—this despite the well-known relationship between trichinosis and garbage-fed pigs. In 1946 postal inspectors in Philadelphia detained three large, foul-smelling packages full of plate scrapings and other food debris. A man named John Wagner, who had mailed the offending parcels to his central Pennsylvania farm, explained that he hated to see good food go to waste, and so had for years been scouring the trash cans behind hotels for provender to fatten his

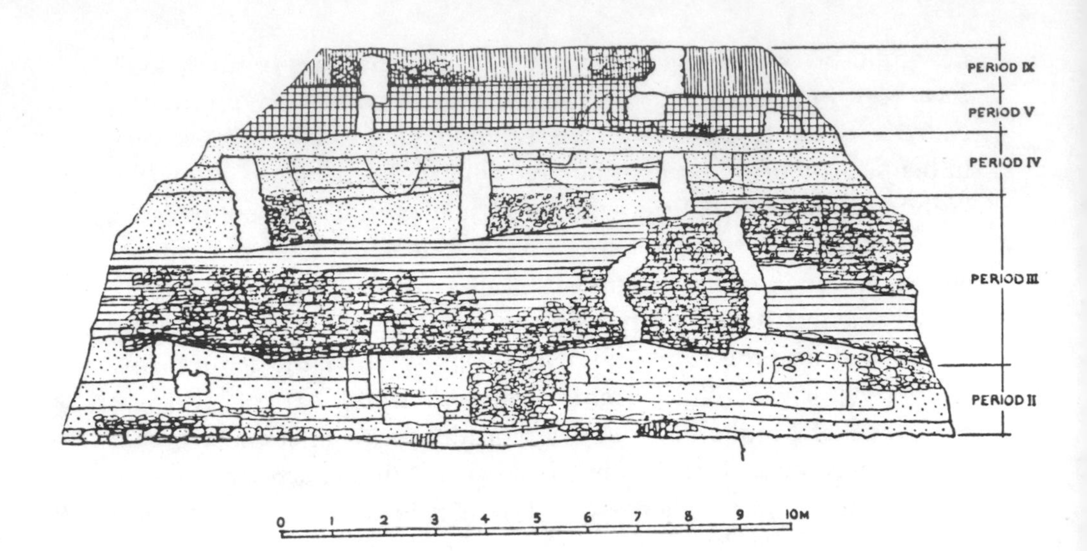

Figure 2-A. A cutaway view of one section of ancient Troy reveals how the city managed, literally, to surmount its garbage problem.

SOURCE: Carl W. Blegen, *Troy and the Trojans* (Praeger, 1963). Reprinted by permission of Greenwood Publishing Group, Westport, CT.

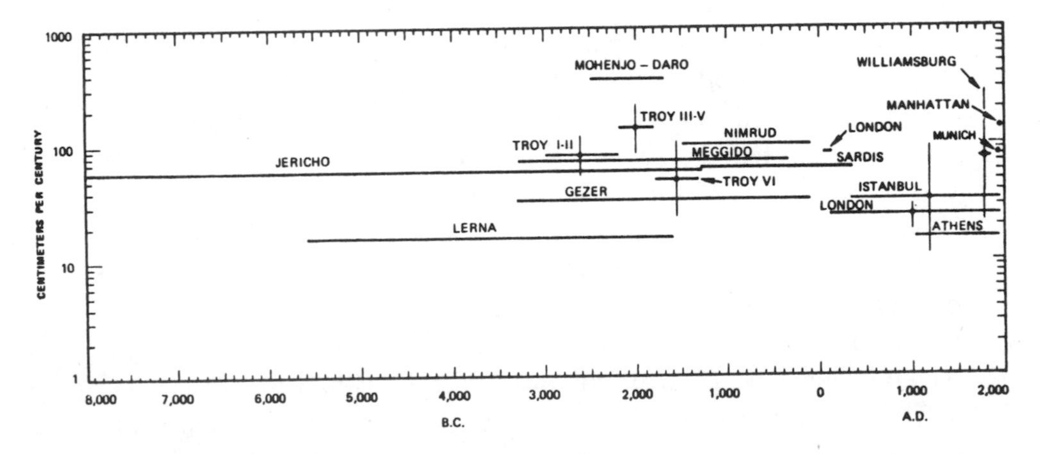

Figure 2-B. A comparison of estimated solid-waste-accumulation rates for various ancient and modern sites.

SOURCE: Charles Gunnerson, "Debris Accumulation," *Journal of the Environmental Engineering Division*, American Society of Civil Engineers, June, 1973

hogs. The slaughter of some 400,000 hogs in the mid-1950s to prevent the spread of a vesicular exanthema epidemic finally moved public-health departments in the United States to prohibit the use of raw garbage as animal feed. It is still legal to use cooked garbage, however, and in various parts of the country, as documented by Orville Schell in his book *Modern Meat,* there are piggeries in which a pig-slopping regime, employing cooked slops, survives.

Needless to say, in ancient times and subsequently, more than food was thrown into the streets. In exchange for the right to sell anything useful that they might find, human scavengers carried much of the inorganic garbage to vacant lots or to the outskirts of a settlement, where it might either be left in piles or burned. In Old Testament times the people of Jerusalem burned some of their garbage in fires emanating from natural gas vents in the nearby Valley of Gehenna, to the south of the city; through a process of association the word Gehenna became a synonym for "hell." If a settlement was occupied for any length of time the piles of refuse, which are known to archaeologists as middens, would naturally become quite large, as layers of newer artifacts were slathered over layers of older ones. Archaeologists realized very early that the strata discernible in middens represented a Rosetta Stone of cultural chronology, and that the information gleaned as a result—about, for example, which styles of pottery were prevalent in what order of succession over time—became a powerful analytical tool. As we have seen, the Garbage Project could use chronological typologies of pull tabs in precisely the same way. The shapes and logo designs on bottles and cans of beer and soda are another useful tool, and they go back much further in time than pull tabs do.

Today, using such technologies as wet-screening and flotation, archaeologists are able to recover a good deal of suggestive organic material from middens, such as tiny fragments or splinters of fish and animal bones, burned seeds, even pollen. But the bulk of what archaeologists find in middens consists of objects made of stone, clay, glass, and metal. Indeed, although "biodegradability" seems sometimes to be held up as a primary characteristic of the garbage of our ancestors—one that, owing to the advent of plastics and other such materials, seems to be increasingly less in evidence today—from the beginning of time nonbiodegradability has been a strikingly constant, even predominant, feature of garbage. Stone tools have remained intact for more than two million years, in every kind of environment. Fired-clay pottery—for cooking, storage, serving, or ceremony—may break into pieces and even discolor a bit, but the pieces themselves are virtually indestructible. Glass is as durable as pottery. Even things that are theoretically biodegradable don't always biodegrade. For example, animal bones decompose in acidic soil, but not all soils are acidic, as is evident from the vast number of bones, human and otherwise, that archaeologists uncover in middens and graves.

Much of the nonbiodegradable matter that turns up in middens, such as intact pottery and utensils, invites speculation as to why it's there in the first place. Those who condemn our own era for its conspicuous consumption and conspicuous waste should at least bear in mind that throwing away perfectly good objects seems to be one of those inexplicable things, like ignoring history, that human beings have always done. David Pendergast, an archaeologist who is a curator at the Royal Ontario Museum, in Toronto, spent seven years studying a Classic Maya site—Altun Ha, in Belize, which was occupied from around 800 B.C. to A.D. 1000—and after examining the contents of various tombs he concluded: "These people would have traded in a Cadillac when the ashtray was full."

How much nonbiodegradable waste did our forebears generate? What proportion of all garbage did it account for? There are, of course, no precise answers to these questions, but certain discoveries give one pause. For example, shell middens—the remnants of countless feasts on clams and oysters by prehistoric Indians—have been discovered by the thousands along the Atlantic coast of North America and along the Gulf of Mexico, and the size of these ancient, unbiodegraded garbage dumps is often startling. There is one, for example, on the Potomac River, at a place called Pope's Creek, Maryland, that covers thirty acres and is an average of ten feet thick. It would take a modern American community of fifty thousand people roughly ten years to fill up an equivalent volume in a landfill. One estimate of the area covered by ancient shell middens in Virginia and Maryland alone is one hundred thousand acres.

Another example from antiquity comes from the results of excavations conducted in Colorado in 1958 and 1960 by the archaeologist Joe Ben Wheat. During late May or early June in the year 6500 B.C. or thereabouts a band of paleo-Indian hunters and their families stampeded a herd of *Bison occidentalis* into an arroyo 140 miles southeast of what is now Denver, at a place known to archaeologists as the Olsen-Chubbuck site. (The time of year in which the event took place can be determined because of the presence of the bones of young bison calves.) Two hundred of the bison were killed, and of these the hunters butchered 150. By one estimate, the hunters carried off enough meat to feed 150 people for some twenty-three days. Behind them they abandoned the leftovers that archaeologists uncovered 8,500 years later: 18,380 pounds of bones. Compare those 18,380 pounds to the total amount of garbage that, according to the highest estimates put forth by the Environmental Protection Agency, 150 latter-day Americans throw away in twenty-three days: a relatively modest 14,145 pounds, which includes all household food debris and food packaging, all nonfood packaging, all yard waste and other household waste, and all the garbage for which these 150 people are responsible in schools, offices, stores, and restaurants. Left in the open, as the bison carcasses were, much of that 14,145 pounds of modern garbage would rapidly biodegrade.

The comparison here is extreme, of course. Most ordinary household waste consists of material that has somehow been processed, and waste is generated at every transformative stage. That waste never shows up in the data on household waste because it gets dealt with somewhere else—at the factory, say, or at the slaughterhouse, or on the farm. Although many of these waste products themselves have further uses—and are not simply discarded—it remains true that Americans are responsible for many times more garbage than the amount they personally throw away. But the Olsen-Chubbuck story draws attention to the fact that garbage that doesn't biodegrade has long been a fact of life. Indeed, because dogs and pigs were available to eat the organic waste that people threw away, and because the eyes and hands of the poor would have been attentive to thrown-away goods that could be reused, nonbiodegradables probably accounted for a very large portion of the garbage that made it to ancient trash heaps. It is only in relatively recent times, with the advent of a civilization that is based on—utterly dependent upon—paper of all kinds, that potentially biodegradable materials have come to constitute a majority of everything that finds its way into a dumping ground.

In most of the Third World a slopping-and-scavenging system that Hector and Aeneas might recognize remains in place. In Egypt the scavengers are known as *zabaline,* and are predominantly Coptic Christians. In Mexico the scavengers are called *pepenadores;* they are unionized and powerful. The image of pestiferous "garbage mountains" in the developing world is at once repellent and almost a cliché, but the people who work these dumps, herding their pigs even as they sort paper from plastic from metal, are performing one of the most thorough jobs of garbage recycling and resource recovery in the world. What's an enlightened, right-thinking environmentalist to say? The garbage mountains are a noisome reminder that a truly efficient system for the disposal of garbage is not always compatible with other desirable social ends—economic development, modernization, and human dignity, for example.

By the same token, the generation of large amounts of garbage—in new and ever-mutating forms—is not necessarily a sign of social woe. When William Stewart Halsted, the chief of surgery at Johns Hopkins University Hospital, became, in 1893, the first surgeon to wear a pair of sterile gloves during an operation—unwittingly setting in motion a chain of events that would turn American hospitals into vast dispensaries of disposable rubber and plastic objects—the goal was not, of course, to create more garbage. It was to make surgery safer for patients. In the United States, a garbage problem is in some respects the price we pay for having learned to do some important things very well.

It was the threat of disease, finally, that made garbage removal at least partially a public responsibility in Europe and the United States. One obstacle these days to a calm and measured approach to garbage problems is a collective memory restricted to the human lifespan of about seventy-five years. It is difficult for anyone alive now to appreciate how appalling, as recently as a century ago, were the conditions of daily life in all of the cities of the Western world, even in the wealthier parts of town. "For thousands of years," Lewis Mumford wrote in *The City in History,* "city dwellers put up with defective, often quite vile, sanitary arrangements, wallowing in rubbish and filth they certainly had the power to remove." The stupefying level of wrack and rejectamenta in one's immediate vicinity that was accepted as normal from prehistory through the Enlightenment was raised horribly by the Industrial Revolution, which drew millions of people into already congested cities and at the same time increased the volume of consumer goods—future throwaways—by many orders of magnitude.

Life magazine fashionably heralded the advent of the "throwaway society" in 1955, but it was a century behind the story. During the late 1960s the archaeologist Daniel Ingersoll undertook the excavation of a waterfront site in the Puddle Dock section of Portsmouth, New Hampshire. The portion of dockage he investigated was built between 1830 and 1840, and over the years a great deal of debris accumulated in and around the adjacent cove. In the 1890s the era of haphazard accumulation of garbage was brought to an end when deliberate efforts were made to fill in the area completely with garbage. Reporting on his findings in 1971 in the journal *Man in the Northeast*, Ingersoll wrote:

> The industrial revolution . . . was supplying the consumer with hundreds of disposable containers and materials by the end of the nineteenth century. The estimated 25,000 cubic yards of fill deposited in the upper portion of Puddle Dock show that the age of the throw-away world began not in the twentieth century but during the nineteenth.

The nineteenth century is the one that gave us tin cans, corrugated cardboard, ready-made clothes, commercial packaging, and factory-cut lumber and other mass-produced construction materials—all familiar constituents of America's landfills to this day. As the historian Martin Melosi has noted in his authoritative book *Garbage in the Cities* (1981), one of the ironies of unbridled laissez-faire capitalism was that it gave rise to a kind of "municipal socialism" as cities were forced to shoulder responsibility for such duties as public safety and sanitation.

Benjamin Franklin instituted the first municipal streetcleaning service in the United States in 1757, in Philadelphia, and it was around this time that American households initiated the practice of digging refuse pits, as opposed to just throwing garbage out of windows and doors. In his book *In Small Things Forgotten* the archaeologist James Deetz sees this newly fastidious behavior in the context of other instances of a late-eighteenth-century craving for order, and he ingeniously ties them all to the waning power of religion and a sense that many aspects of life were increasingly out of control. Be that as it may, a recognizably modern approach to urban sanitation had to await the late nineteenth century and the pioneering efforts of Colonel George E. Waring, Jr., "the Apostle of Cleanliness." Waring, a Civil War veteran and a protégé of the landscape architect Frederick Law Olmsted, was named to the position of Street Cleaning Commissioner of the City of New York in 1895, during one of New York's periodic spasms of reform, and he set up the first comprehensive system of public-sector garbage management in the country. Waring and his two thousand white-clad employees—they were known as the "White Wings"—cleared the streets of rubbish and offal and carted off their cullings to dumps, incinerators (known then as "cremators"), and, until the affluent owners of shorefront property in New Jersey complained, to the Atlantic Ocean. Some garbage was subject to "reduction," a technique imported from Europe in which wet garbage and dead animals were stewed in large vats in order to retrieve various byproducts.

Although Waring was ousted by a revitalized Tammany Hall in 1898, his powerful image as the commander of legions and protector of the public health influenced communities everywhere. In 1880, according to data gathered by the historian Melosi, fewer than a quarter of America's cities could boast a municipally run system for disposing of garbage. By 1910, eight cities out of ten could. Though it is often forgotten amid the well-publicized worries about our present situation, taking the long view generally brings home the fact that ever since governments began facing up to their responsibilities, the story of the garbage problem in the industrialized world has been one of steady amelioration, of bad giving way to less bad and eventually to not quite so bad. To be able to complain about the garbage problems that persist—and, indeed, to harbor the hope, even the expectation, that they will one day, somehow, be addressed, though that day may not be tomorrow—is yet one more luxury that Americans are unaware they enjoy.

The advent of systematic waste collection did not put an end to scavengers or to the significant recycling function that scavengers performed, but it did decisively shift the locus of scavenging from the personal level (in many places it was a familiar and accepted feature of daily life) to the commercial. As the operations at dumps and landfills grew increasingly vast and mechanized, the presence of ordinary people became a nuisance (and an invitation to lawsuits stemming from injuries). The owners of disposal sites began declaring their properties off limits to casual scavenging, thereby helping to put an end in many parts of the country to a widespread social and economic ritual: the Sunday afternoon excursion to drop off the family's garbage and perhaps pick up some gossip and a discarded item or two. In a famous essay in *The Atlantic Monthly* in 1959 Wallace Stegner recalled the town dump in his youth (in Whitemud, Saskatchewan), observing that "it contained relics of every individual who had ever lived there, and of every phase of the town's history." He went on to rhapsodize about the abandoned bedsprings, the old books, the broken dishes and rusty spoons:

> There were also old iron, old brass, for which we hunted assiduously, by night conning junkmen's catalogues and the pages of the *Enterprise* to find out how much wartime value there might be in the geared insides of clocks or in a pound of tea lead [used in the lining of tea chests] carefully wrapped in a ball whose weight astonished and delighted us. Sometimes the unimaginable outside world reached in and laid a finger on us. I recall that, aged no more than seven, I wrote a St. Louis junk house asking if they preferred their tea lead and tinfoil wrapped in balls, or whether they would rather have it pressed flat in sheets, and I got back a typewritten letter in a window envelope instructing me that they would be happy to have it any way that was convenient for me. They added that they valued my business and were mine very truly.

The kind of enchantment that Stegner evokes can still be encountered in rural (and, for the most part, illegal) dumps, where the right to scavenge is almost as sacred as the right to bear arms. But it has virtually vanished everywhere else, and landfill owners are vigilant. According to the *Philadelphia Inquirer*, in December, 1979, just before Christmas, in Cheraw, South Carolina, a garbageman named

Raymond Sandsberry, Jr., the father of seven children, was arrested for removing from a landfill nine pairs of shoes, forty items of clothing, and a woman's handbag. Incidents like that one have had a chilling effect on individual scavenging.

In contrast, corporate scavenging—the retrieval and marketing of what are known as "secondary materials" (scrap metal, for instance: everything from junk cars and trucks to "white goods" such as used refrigerators and stoves)—remains a big business to this day, with its own trade associations (the most prominent is the Institute of Scrap Recycling Industries, known as ISRI), its own annual conventions, its own distinct sociology. At a time of chronic U.S. trade deficits, scrap metal is a significant American money-maker, accounting for three-quarters of all the ocean-borne bulk cargo that leaves the Port of New York and New Jersey—1.6 million long tons a year, most of it bound for Korea, India, and Taiwan. At the turn of the century, though, the commodity of choice for professional garbage scavengers was rags, which were used in the manufacture both of low-cost garments and of paper. The collection of newsprint and cardboard for recycling was also considerable. Without the slightest encouragement from Friends of the Earth or the Committee for a Better Tomorrow—which, like most environmental organizations, did not exist at the time—a significant portion of discarded rags and paper was being recycled in the United States in the early 1900s. The reason was economic. Measured in 1990 dollars, the price per ton of rags was $350, which is not much below what aluminum, one of the most lucrative of modern recyclables, fetches today. A ton of waste paper could be sold for about $160, which in many places is about $160 more than waste paper can be sold for now. New York State, which at the time was the nation's largest producer of newsprint, recycled almost 15 percent of its waste paper in 1900. By the 1920s, however, wood-processing technology had matured and rail links to the forests of the Northwest had been secured, and wood replaced rags and used paper as a fiber source. The rag trade was dealt a final blow by the Wool Products Labeling Act of 1939, which required products made out of recycled wool and cotton fibers to be so labeled, the effect of which was to devalue products by implying inferior quality. The implication did not need belaboring. Woolens made at least in part out of previously used wool had been known as "shoddy," and this noun quickly evolved into an adjective with pejorative connotations.

The vagaries of the secondary-materials markets underscore yet another fundamental garbage reality: desirable things happen to garbage mainly when someone stands to earn money by making desirable things happen. Good intentions alone don't count for much. Despite what people profess in opinion polls as to what they would "be willing" to do with their garbage or what they would "be willing" to pay, the truth is that high-mindedness often stops at the garbage can's rim.

A century of avid, painstaking archaeology on six continents by thousands of scholars has yielded tome after tome in which the secrets held by ancient discards have been revealed. The contents of the household garbage of our own time, in contrast, remain largely a mystery. Americans are, admittedly, exquisitely sensitized to the existence of something called "litter," and they know full well the nature of the objects that litter tends to consist of. But litter makes up an infinitesimal fraction of the garbage that this country produces, and Americans don't have a clear idea at all about what the garbage that isn't litter actually contains.

This should not really be surprising. Unlike the evidence of many another problem, be it a social one, such as poverty, or an aesthetic one, such as bad architecture, the evidence of specific pieces of household garbage disappears from one day to the next. People put their garbage in the garbage can under the kitchen sink, in the bathroom, in the den, and then someone collects it all and *takes it out.* The garbage that is taken out is eventually left at the curb or in the alley, and very soon it is gone. All of this garbage is quickly replaced by other garbage. Garbage passes under our eyes virtually unnoticed, the continual turnover inhibiting perception. One of the handful of things that every American does every day—throw garbage away—is among the least likely of all acts to register. The cliché about garbage we've all heard is: "Out of sight, out of mind." Yet even when it's *in* sight garbage somehow manages to remain out of mind.

That individual failure to perceive has its counterpart in American society at large. No one in this country really knows how much garbage Americans produce. No one knows what kinds of hazards we will face in decades to come from the garbage, some of it toxic, that is already buried in the ground. And no one really has a firm grasp on the totality of behaviors—what archaeologists call "formation processes"—that result in the creation and discard of this kind of garbage or that. Ignorance is one of the biggest handicaps we face when it comes to deciding, as a society, whether and how to throw various kinds of garbage away.

What we have so far been calling "garbage" sanitation professionals call solid waste. There are many categories of solid waste, and the most significant categories of all—those associated with manufacturing, mining, agriculture, and so on—are the ones we tend to think least about, although together they constitute more than 98 percent of the twelve billion tons of material in America that in some sense get discarded every year. The solid waste we're all most familiar with—the kind on which present-day concerns about garbage disposal are centered, and the kind on which the Garbage Project has concentrated its attention—is the solid waste that comes from the households and institutions and small businesses of towns and cities: "municipal solid waste" (MSW). Professionals talk about the municipal solid waste that we throw away as entering the "solid-waste stream," and the term is an apt figure of speech. Waste flows unceasingly, fed by hundreds of millions of tributaries. While many workaday activities come to a halt on weekends and holidays, garbage flows on. Indeed, days of rest tend to result in the largest waves of garbage. Christmas is a solid-waste tsunami.

It stands to reason that something for which professionals have a technical term of long standing—"solid-waste stream"—should also have, if nothing else, a weight and volume associated with it, but "stands to reason" is a phrase that all too frequently augurs a wrong turn. In this case the fact is that estimates of the amount of garbage produced in the United States every day (or year) vary so widely as to be useful only up to a point. There has, nonetheless, been a great deal of vivid imagery. Katie Kelly, in her book *Garbage* (1973), stated that the amount of municipal solid waste produced in the United States annually would fill five million trucks; these, "placed end to end, would stretch around the world twice." In 1988 *Newsday* cited a New York State legislative committee estimate that a year's worth of America's solid waste would fill the twin towers of 187 World Trade Centers. That same year the Orlando *Sentinel* estimated that the total annual volume of U.S. solid waste would cover the entire 43,600 miles of the interstate-highway system to a depth of seven and a half inches. *The Baltimore Sun* recently claimed that Baltimore generates enough garbage every day to fill Orioles' Stadium to a depth of nine feet—a ballpark figure if ever there was one.

Information of this kind is unreliable, and its origins not a little mystifying. Because virtually all the data that exist on solid-waste quantity were, until very recently, rendered in terms of weight, not volume, one has to wonder how weight data were converted to volume data for Kelly's trucks, *Newsday*'s towers, the *Sentinel*'s highways, and *The Sun*'s stadium. Rough ratios comparing the percentage of landfill contents that various garbage components take up in terms of volume and in terms of weight—plastic (2.5:1), paper (1.1:1), metal (1.6:1), glass (0.2:1), food (0.4:1), and so on—were worked out only in 1990 in a joint effort involving the Garbage Project and the environmental consulting firm Franklin Associates. These would enable us to make a weight-to-volume conversion—taking into account the consequences of compaction, which, as we will see, are severe—for a landfill whose moisture content was average and whose proportional make-up, by type of garbage, was known. But a universal volume-to-weight ratio covering all garbage everywhere is probably impossible to devise. Among other things, the very same types of garbage vary enormously in both weight and volume from place to place. Thanks to rainfall and high humidity, for example, a bag of garbage from New Orleans will at certain times of year weigh half again as much as a bag of exactly the same garbage from New York City. Even if an overall volume-to-weight ratio for garbage did exist, there could still be biases involving weight to distort the calculations. In the years before the First World War it was widely suspected that the weight of the garbage being trucked away from cities was sometimes being inflated in order to bring home to politicians the need for greater sanitation efforts—a mild sin, perhaps, in the service of the common weal. Because payments for carting away garbage and for dumping ("tipping") at landfills are most frequently based on tonnage, there was also long rumored to be a parallel tendency among some haulers and landfill managers to err on the high side in their daily tasks.

All of the above aside, what is the word "volume" being taken to mean? Is it the volume of garbage "as discarded" in garbage cans? Is it the volume of garbage as it arrives at the landfill—a fraction of its former size, crushed under a pressure of fifty-two pounds per square inch by the hydraulic ram of a standard "mother hen" compactor truck? Is it the even-further-compressed mass that is squeezed into an endloader-rolloff at a transfer station? Or the smaller-still volume that results after garbage has been buried for years and years under tons of other garbage in a landfill? The figures that are tossed around almost never make the answer clear. And yet garbage gets so compressed that a cubic yard's worth of it, which might weigh 100 pounds tumbling fresh from the can, turns into a dense package weighing anywhere from 800 to 1,400 pounds by the time it is deposited in its final resting place.

There have been, to be sure, some careful, professional attempts over the years to determine the total amount of municipal solid waste that Americans throw away in a year, though these, too, suffer from flaws. One group of investigations was conducted during the mid-1970s, shortly after the Environmental Protection Agency was created. Another series of investigations has been undertaken within the past five years, in response to a decline in the number of operating landfills. Calculating a figure for the amount of garbage produced annually in the United States is a daunting task if for no other reason than that one cannot, of course, weigh or examine more than a tiny fraction of the whole. All serious studies have had to take short cuts. Some have tried to capture an accurate picture of disposal patterns in, say, twenty towns and cities, and have then gone on to extrapolate the findings to the nation as a whole. Other studies have used what is called the "materials-flows method" to estimate garbage generation, although, ironically, this method doesn't involve examining garbage per se at all, but rather examining industrial-production and -consumption records—how much of everything is being made and used. To these data are applied certain assumptions about American discard patterns, and the result is an estimate of the rate at which materials are entering the solid-waste stream.

The materials-flows method is ingenious, but one whose utility is somewhat undermined by the fact that many of its assumptions are untested. One materials-flows study, for example, assumed that the useful life of major household appliances is twenty years, after which time it was assumed that the appliances would be thrown away. That assumption ignores (as we will see in chapter nine) the substantial underground trade in used durables that supplies many low-income households with washing machines and refrigerators and the like, and that is also a source of parts no longer carried by local dealers. The materials-flows method also takes insufficient account of the collection and export of scrap iron and steel. It is not within the method's power to do better than guess (based on extrapolations from a handful of local studies) at the amounts of food and yard waste that get tossed into the garbage. And the materials-flows studies of landfill contents don't take into account construction and demolition debris, which technically (according to the Environmental Protection Agency's definition) doesn't count as municipal solid waste, but a lot of which ends up in municipal landfills.

Not surprisingly, then, the estimates of the size of the U.S. solid-waste stream range widely. Generation-rate figures are most commonly expressed in pounds discarded per person per day, and various studies from the past decade and a half have arrived at the following rates for municipal solid waste: 2.9 pounds per person per day, 3.02 pounds, 4.24, 4.28, 5.0, and 8.0. (For the record, the figure for the people who killed the Olsen-Chubbuck bison, based on the weight of the discarded bones alone, is about 5.3 pounds.) The two most comparable studies that have been conducted in recent years have been materials-flows studies done for the EPA by, in one case, the agency's Office of Solid Waste Management Programs, in 1977, in conjunction with the consulting firm Franklin Associates, and in the other, in 1986, by Franklin Associates alone. These used the same methodology and the same database, but several of the assumptions employed in the earlier study were revised in the later one. The result is that, for the years covered by both, the later study found garbage production to have been 20 percent less by weight than the earlier study did. All told, estimates of the amount of garbage we generate, individually or as a nation, leave a lot to be desired —a good argument for requiring local communities to conduct regular, standardized waste inventories.

But what about the question to which we tend to assume the answer must be yes: Are Americans, on a per capita basis, bringing into existence a lot more municipal solid waste than they did twenty, fifty, or a hundred tears ago? For the reasons we have just discussed the question cannot be answered with precision, but the answer all the same may very well be no. As one might imagine, not very much comparable data is available on garbage-generation rates during different periods of time, but what little there is does not support the view that per capita rates have greatly accelerated over the years. Garbage Project sorts of large amounts of purely household garbage in Milwaukee during the late 1970s found that households there threw out garbage at a rate of about a pound-and-a-half per person per day. Fortuitously, data exist for Milwaukee from a period twenty years earlier—1959, specifically. A study done at the time for a doctoral dissertation by John Bell of Purdue University found that Milwaukee households were throwing away slightly more garbage than the Garbage Project would find: about 1.9 pounds per person per day. Admittedly, these data involve only household waste, not the larger category of municipal solid waste. But household waste is by far the largest category of MSW, and the Milwaukee comparison at least deserves a place in the evidence pile.

Looking at the matter another way, let us assume to be correct the Environmental Protection Agency's estimate (probably too high) that the average American throws out about fifteen hundred pounds of garbage a year. That certainly seems like a lot. History reminds us, however, that many former components of American garbage no longer exist—major components, whose absence does not even register in the collective memory. Thus, we do not see the twelve hundred pounds per year of coal ash that the average American generated from home stoves and furnaces at the turn of the century, and that was usually dumped on the poor side of town. We do not see the more than twenty pounds of manure that each of the more than three million horses living in cities produced every day at the turn of the century, or the hundreds of thousands of dead city horses that once had to be disposed of every year. We do not see all the food that households once wasted willy-nilly because refrigeration and sophisticated packaging were not yet widespread.

Several points should be emphasized. First, some of what used to be household waste, such as coal ash, is now produced by public utilities, and does not become part of municipal solid waste—but it is still waste, obviously. Coal-fired utilities, though, provide less than a quarter of America's electricity; the ash they produce is disposed of by the utility (usually on site) and is not a culprit in the rapid filling of municipal landfills—the phenomenon that, more than any other, initially provoked garbage-crisis fears. Second, it is undeniable that Americans as a whole are producing more municipal solid waste than they did fifty or a hundred years ago; it should be understood, though, that this is largely because there are more Americans than there were fifty or a hundred years ago. Debates can and do swirl these days about per capita generation rates, and whether they've been going up slightly year by year in recent decades, and by how much, if any. Certainly wars, recessions, and social innovation (for example, the advent of curbside recycling) wreak annual variation on the solid-waste stream, though in ways that economists and social scientists cannot yet successfully describe. But a long view of America's municipal solid waste would suggest that, on a per capita basis, the nation's record is hardly one of unrestrained excess. Indeed, the word that best describes the situation with respect to overall volume may be: stability.

There is a new, radical branch of archaeology—it is called "critical archaeology"—which reminds us of the fact that our own latter-day attitudes, together with the objects with which we're familiar and the techniques we employ to acquire and disseminate knowledge, inevitably introduce a bias into our reconstructions of the past. Sometimes the biases are subtle and treacherous. Sometimes they are crude and readily visible. On the cover of tourist brochures, the tall temples and palaces of the ancient Maya stand limestone-white against the green of the surrounding forest canopy, and in our imaginations we think of them as looking this way in bygone ages, although in fact large parts of them once were painted in bright reds and yellows and blues. Similarly, we think of the Parthenon in Athens as having always stood in blinding purity beneath the azure Mediterranean sky, although it, too, was once rendered garishly. We think of a visit to colonial Williamsburg as a walk back in time—

"where the eighteenth century still lives," the advertisements say—forgetting about the lawnmowers and weed-eaters that have taken the place of chomping cows and sheep; about the nonpeel, water-seal paint that keeps the buildings in a state of unwonted tidiness; about the asphalt that now lies where once were rutted, muddy, manure-laden tracks; about the garbage trucks that rumble through as often as three times a day, carting away what in former times would have festered in redolent piles. A Williamsburg that offered a real taste of eighteenth-century life would be closed down swiftly by public-health officials.

The most extreme among the critical archaeologists would hold that the past can exist *only* as a reflection of the present—much as the ship in the AT&T commercial existed as a reflection not of what it really was but of what the commercial's creators wanted it to be. That is going too far. One can stop well short of this position and yet agree that historical reality presents limits to our kenning—and that the present is a distorting lens through which we have no choice but to look. The ignorance and misconceptions about garbage in history serve as a case in point—and as an object lesson.

CHAPTER 3

WHAT WE SAY, WHAT WE DO

When Thomas Price took command of the Sanitation Division of the City of Tucson, in 1966, he found a department whose members suffered from high rates of alcoholism and high rates of absenteeism, an uneviable safety record, and exceedingly low morale. Price, a bulky, convival, immensely competent man, and one who shared the Hispanic roots of most of his employees, focused first on morale. From the University of Arizona's film library he obtained documentaries about the links, via rodents and insects, between uncollected garbage and infectious disease, and he showed these films week after week to remind his workers that they were not simply clock-punchers but agents of public safety. He warned them again and again about the dangers inherent in their work—from microbes and toxic waste in household discards, to some extent, but mostly from the heavy machinery that is involved in every stage of the garbage-disposal process. (According to the Bureau of Labor Statistics, the incidence of occupational injury among sanitation workers in 1986 was 177 injuries per 1,000 workers, compared with an average that year of 77 per 1,000 workers in the entire private-sector labor force; a study in New York City covering the period 1973–1983 found the "injury-severity ratio" among municipal sanitation workers—that is, the number of days lost by sanitation workers per incident of injury—to be equivalent to that among mineworkers.) And Price ordered workers with alcohol problems to get treatment or get out. He built a palpable esprit de corps within the Sanitation Division, and before long there was a waiting list of applicants for jobs in garbage disposal. Price's reward came when, in 1973, he was named the director of Tucson's entire Department of Operations.

Tom Price played a key role in the founding of the Garbage Project. When representatives of the University of Arizona's anthropology department met with him to discuss their plans to mount a study of Tucson's garbage, and to ask for his help, they found not the hidebound bureaucrat they feared—rigid, myopic, obstructionist—but rather an enlightened despot, a philosopher-garbageman (and a University of Arizona alumnus). "Why not?" was his response when the request to collect garbage for study was made. "People threw it out, didn't they?" And thus it came to pass, in the spring of 1973, that the first teams of anthropology students settled in behind a row of dumpsters at the Sanitation Division maintenance yard, on South Tenth Street, where four days a week sanitation supervisors stopped by in pickup trucks to deposit fresh garbage for analysis. Tom Price, who died of leukemia in 1988, at the age of 57, was honored from the outset by Garbage Project personnel as "Santo Tomás."

Price seemed to grasp instinctively one of the central tenets of the Garbage Project: that what people have owned—and thrown away—can speak more eloquently, informatively, and truthfully about the lives they lead than they themselves ever may. People such as Price, who work with garbage on a daily basis, seem to come to that conclusion naturally. In the early 1970s, a garbageman named Frenchy Benguerel, of Kenwood, California, made the same point during an interview with Charles Kuralt for the "On the Road" segment of the "CBS Evening News." "Can you tell a lot about the customers from their garbage?," Kuralt asked Benguerel. "Oh, definitely," Benguerel replied. And he went on:

> You can tell what kind of wine they drink. All their letters come in and out, and who they buy through—Saks or Sears and Roebuck—and how they maintain their household. It's better'n being a psychiatrist. I can tell you anything you want to know.

The assumption that behavior is reflected in artifacts—and, depending on the situation, in the lack of artifacts—lies at the heart of studies of what is known to archaeologists as "material culture." Students of material culture think of physical artifacts (from the garbage in our waste baskets to the paintings on our walls) as not only helping to define us at any given moment but also as contributing to a changing of the definition itself over the course of time. A highway does not merely reflect a static pattern of traffic; it transforms the vectors of building and development. Microwave dinners and McDonald's hamburgers do not merely reflect a new diversity in work and family; they contribute to that diversity.

Modern material-culture studies, which were prefigured in a way by pop art's apothesis of the commercial and the mundane, have become a recognized and legitimate research endeavor in a variety of scholarly disciplines. Archaeologists, of course, have been picking over material culture's leftovers for years: In too many cases, those are the only clues to past behavior that archaeologists have had. But now environmental psychologists, architects, and urban planners are studying the impact of the material environment on behaviors and attitudes (and vice versa). Market researchers and consumer educators are focusing on the interrelationships among commodities, attitudes, and behaviors, because those relationships can be a key to the efficient selling of products and concepts. Sociologists are showing interest in material culture, because it can sometimes offer a way of corroborating and correcting information obtained in interviews; a way, that is to say, to circumvent the problem of "informant bias." Indeed, within many disciplines these days, material-culture studies are viewed as an essential adjunct to studies based on interviews and surveys.

The Garbage Project, being largely though not exclusively an archaeological endeavor, pursues the aims and is heavily reliant on the techniques of material-culture studies. As noted in chapter one, the

Project grew out of an anthropology course in which students were focused precisely on the question of how to discern links between physical evidence, often fragmentary, on the one hand, and mental attitudes and patterns of behavior on the other. One elementary but memorable study of this kind was done by a student in the course named John W. Hohmann, who investigated the last reaches of a secluded, dirt-packed spur off Trail's End Road in the Sonoran desert northwest of Tucson. For purposes of analysis Hohmann imposed a grid system over a 200-foot by 250-foot area. He then conducted an inventory of the site, recording on his map the distributions of glass scatters from hundreds of broken bottles (mostly beer bottles), and the locations of 133 cans (mostly beer cans), 27 "sex objects" (mostly used condoms), 212 facial tissues, 15 articles of clothing (mostly men's and women's underwear), and 11 "trash objects" (mostly sex or movie-star magazines). Hohmann's map of the site (see Figure 3-A) produced a vividly clear picture of sexual and drinking activity conducted primarily inside of cars (no artifacts were found on the roadway itself), with fragments of glass bottles clustered close by the roadside (where the bottles had been thrown no doubt so that they could be observed to break) and cans in a perimeter farther beyond, and with sexual activities concentrated in the section of the turnoff road best hidden from the main road by a rocky knoll. Subsequent studies of a similar nature found much the same pattern at other road-end locations. John Hohmann, by the way, continued to pursue a career in archaeology. He was the leader of the archaeological team whose recent investigations of the ancient but largely overlooked Casa Malpais pueblo, near Springerville, Arizona, led to the discovery of a system of underground catacombs that had been used by a settlement of Mogollon people (a prehistoric group that vanished mysteriously in the mid-fifteenth century). The catacombs are the first to have been found north of Mexico.

While practitioners of material-culture studies tend to assume, at least in theory, that a dynamic relationship exists between artifacts on the one hand and attitudes and behaviors on the other, the precise nature of the relationship cannot always be stated, and similar patterns of physical evidence do not always indicate similar patterns of behavior. One illustration of the pitfalls involved is provided by an odd discovery made early by Garbage Project sorters—namely, that

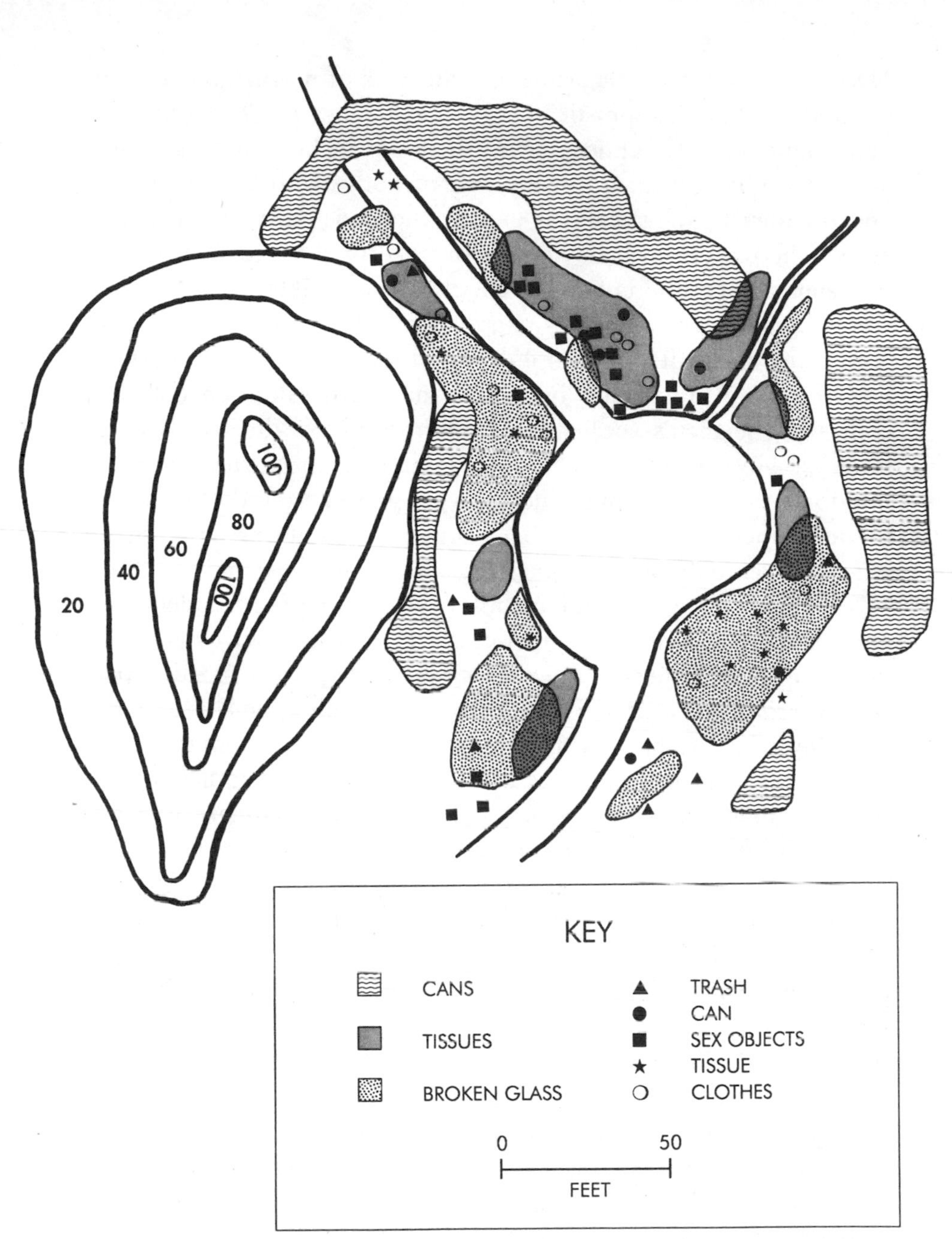

Figure 3-A. Distribution of artifacts at the Trail's End site, outside Tucson. The main road, not visible, runs to the left of the knoll.
SOURCE: The Garbage Project

National Geographic magazine and the kind of magazine known in the trade as "men's sophisticates" (magazines like *Playboy* and *Penthouse* and the rest) almost never show up in household garbage. While the physical record in the case of both *National Geographic* and the men's sophisticates is exactly the same—that is, no magazines—the pattern of behavior responsible for that record is probably somewhat different in each case. Subscribers to *National Geographic* tend to keep the magazines on the shelf for a long time—in many cases, literally, until death (at which point the magazines find their way into other homes or into yard sales). In contrast, copies of the men's sophisticates often don't get thrown away at home because people, out of embarrassment, throw them away in nondomestic garbage cans. They also get passed around—from, say, home to locker room to barber shop to gas station—and, again, wind up eventually in nondomestic garbage cans.

Still, in the face of all the complexities, and perhaps understandably, some researchers in practice have typically viewed material objects not as partners in a dynamic relationship but instead as passive reflections of attitudes and behaviors. Building on this assumption, investigators have worked to establish correlates that seek to link specific physical evidence or a specific physical milieu with a specific mindset or with specific habits and other personal characteristics. For example, a number of studies have tried to turn the type of house, furniture, clothing, and other such things that people own into indicators of social status or other ineffable qualities. (A typical result is a study like "Living-room styles and social attributes: the patterning of material artifacts in a modern urban community," by E. O. Laumann and J. S. House; they concluded that "people with traditional decor are also more traditional in their behavior and attitudes.") Simple studies of this kind by Garbage Project students have, among other things, attempted to correlate income level with the absence or presence of lawn art, such as plaster fawns or elves; the use of hallucinogenic drugs with odd bathroom decoration (such as black-light posters); and sexist attitudes in different communities with the size of male and female gravestones. The correlations did exist in all cases, but were weaker than might be supposed.

These are studies in search of stereotypes, and they're fine as far as they go. The basic goal is to construct a neatly organized world in which a certain specific aspect of material culture is isomorphic, to use an academic term that simply means "congruent," with certain specific attitudes and behaviors. Such research is often frustrated when a specific type of material culture does not fit into neat patterns with behavior or attitudes. When that happens, one of two conclusions, or both, can be drawn: 1) our stereotypes may be too simplistic; 2) our correlates may be variable because some parts of society are in a state of flux, and so is, therefore, the relationship between the material and the behavioral. From the very outset Garbage Project researchers bore these conclusions in mind and tried to heed their implications.

That first year of the Garbage Project was one of discoveries large and small. The garbage itself was an unknown world—everything learned about it was new—and thus held the fascination that a trip up the Congo in the nineteenth century would have. One of the first discoveries was simply that a substance to which the term "slops" was applied congregates at the bottom of every paper or plastic bag into which garbage is dropped. Slops (Garbage Project code number 069) comprise a stew of such things as coffee grounds, fruit parts, rotten vegetable bits, cigarette butts, grit of unknown origin, and the sort of gooey canned mush epitomized by Chef Boyardee ravioli; somehow, in the course of every garbage bag's journey from kitchen to truck, all of these substances find one another and intimately coalesce. The Project eventually undertook a detailed investigation of the tiny individual constituents of slops, which, based on refuse pickups from sixty-nine households in seven census tracts, were found to consist primarily of bakery products and cereal (28 percent by weight); fresh vegetable matter (24 percent); high-protein vegetables (12 percent); meat, poultry, and seafood parts (8 percent); fruit waste (8 percent); cheese and other milk products (6 percent); and fats and oils (5 percent). Most slops originate in the form of plate scrapings; the reputation of vegetables as prime candidates to become leftovers appears to be well deserved.

Another phenomenon that quickly became clear was the capacity of garbage to surprise. This was vividly brought home to researchers as the result of the discovery, by an anthropology student named

Diane Tucker, of a diamond ring amid a mass of potato peels. (The ring, a relatively inexpensive one, could not be returned, because of Garbage Project procedures to ensure that the identity of the households from which garbage for study is obtained remains unknown; it was accidentally thrown away along with other prospective exhibits for a Garbage Project museum, all of which had been stored in a special dumpster.) Most of the surprises, however, have not been so immediately obvious. They have not, in other words, tended to be the garbage equivalent of finding the Mask of Agamemnon or the cave paintings at Lascaux. Rather, they have emerged through the careful recording of each and every artifact found in each and every load of garbage, and the statistical evaluation of the results.

A good example that comes from the Garbage Project's first two seasons involves red meat. The counterintuitive nature of the findings are typical of what garbology frequently turns up. During the spring of 1973 there had been a widely publicized beef shortage in the United States. From March through September a good selection of beef in supermarkets was hard to find, and the meat was very expensive. The Garbage Project, which from the beginning has been very interested in food waste, decided to look into discard patterns of red meat to see if people's behavior changed appreciably between times of shortage and (afterwards) times of plenty. As it happens, meat is an ideal subject for investigation, because supermarket meat-counter packages are labeled with the type of cut, the weight, the price, and the date of packaging (which is usually on or very near the date of sale); it is thus possible to compare the amount of wasted meat thrown away in garbage with the amount of meat that was originally bought.

Garbage data on beef were collected over a period of fifteen months, from the spring of 1973 through the spring of 1974, and the numbers, when analyzed, revealed a strange pattern. In the months after the beef shortage ended, the rate of beef waste (cooked and uncooked, but not counting fat or bone) amounted to about 3 percent of all the beef bought. During the months of the shortage, in contrast, the rate of waste was 9 percent. In other words, people wasted three times more beef when it was in short supply than they did when it was plentiful. This conclusion seemed perverse, but the data, when checked, seemed solid. Eventually a hypothesis was put forward to account for the odd behavior: the practice of crisis-buying. When confronted with the widespread and sometimes alarmist coverage of the beef shortage in the local and national media many people may have responded by buying up all the beef they could get their hands on, even if some of the cuts were unfamiliar. Of course, they didn't necessarily know how to cook some of those cuts in an appetizing way. More important, they didn't necessarily know how to store large amounts of meat for an extended period of time. The inevitable result in either case: greater waste. This hypothesis was buttressed by the discovery in some Tucson garbage samples, during the shortage months, of a few whole cuts of beef, which it is very likely had spoiled prior to being discarded.*

The general proposition drawn from the findings about red meat —that wastage of a food increases when that food is scarce—was unexpected, but in the context it seemed reasonable. The reaction among nutrition educators and home economists when this result was reported, however, was somewhat muted, their criticism being that the hypothesis was probably not broadly applicable to a wide range of foods. Fate smiled on the Garbage Project in the spring of 1975 by unleashing a sugar shortage. As the price of sugar and high-sugar products doubled, the wastage of those items in Tucson's garbage tripled. Because Tucson is only sixty miles from the U.S. border with Mexico, where the price of sugar had remained stable, many Tucsonans stocked up with sugar that they bought south of the border. Mexican sugar, however, is not as highly processed as American sugar; it is browner, and it turns hard quickly. Before long, hard, brown bricks of Mexican sugar began appearing in the garbage. Some Tucsonans began buying Desserta and other unfamiliar

* Further circumstantial support unexpectedly came some years later from outside of Tucson during the excavation of the Mallard North Landfill, near Chicago. As a sample of landfill waste was being emptied onto a sorting table, one of the sorters suddenly held up a large, dull-brown mass speckled with pale-white blobs and ringed by a pink-white rind. The object was a steak. A question was raised: Could the discarded steak be dated? The other contents of the sample were quickly sifted for pieces of newspaper. A page was found with the date "April 23," but the year was torn off. More garbage was sifted, and another page was found: "May 5, 1973." Then another and another, with dates from the same period. The sample had been plucked from a landfill stratum that coincided with the national beef shortage.

products made from sugar substitutes, such as cyclamates; the reviews were plainly evident in the form of unconsumed discards. Also prominent in the trash were items containing sugar that had crystallized during the course of long-term hoarding. In sum, the behavior of people in the midst of the sugar shortage corroborated the findings about red meat. The sugar shortage, more sharply than the beef shortage, also drew attention to the role that unfamiliarity with a food plays in the wasting of that food.

From the information garnered during the beef and sugar shortages the Garbage Project developed the First Principle of Food Waste: *The more repetitive your diet—the more you eat the same things day after day—the less food you waste.* In hindsight the First Principle seems simple and obvious. The waste in garbage from the standard sixteen-ounce and twenty-four-ounce loaves of sliced bread that every household buys regularly is virtually nonexistent—at most, crusts and ends; this is because common sandwich bread is used continually, meal after meal. But specialty breads—hot dog buns, bagels, muffins, biscuits, kaiser rolls—are wasted at rates of 30 to 60 percent, because they are bought less regularly and perhaps used once or twice in very specific kinds of meals before finding a place in the bread box or the back of the refrigerator to harden or decay. The First Principle helps to explain why the garbage collected from Mexican-American census tracts generally has less food waste—sometimes more than 20 percent less—than does garbage from Anglo census tracts. Mexican-American border cuisine offers a diverse array of dishes, but the ingredients are few: tortillas; beans; chunks of beef, chicken, and pork; avocados, tomatoes, lettuce, onions; red and green chili sauce; salsa. Not only is it easy to incorporate leftovers into new meals, but the staple ingredients are in a state of constant turnover.

No one is very happy about the First Principle. Nutritionists are always trying to get people to broaden their dietary horizons. Producers and marketers of food and other goods make a significant part of their living by providing novelty and diversity, and those very qualities are, as it happens, deeply appreciated by consumers. Still, gaining insight into certain basic elements of the architecture of behavior can have positive practical consequences—in this case, with respect to the reduction of food waste. Garbage Project studies indicate that American families waste between 10 and 15 percent of the food they buy. In the history of the Project, of all the bags of household garbage that have been examined, only a handful have been found to be innocent of wasted food.

The Garbage Project moved from the Sanitation Division's Tenth Street maintenance yard to a site on the University of Arizona campus in 1984. Thomas Price, who was the head of the university's Hispanic-alumni association, and was well-acquainted with members of the university's board of regents, had used his gifts of persuasion to the utmost. In the lot where the Project's staff members now sort garbage there is a trailer to store equipment—scales, aprons, gloves, masks, recording forms, and so on. A ramada has also been built there, with a corrugated metal roof; two of its sides are defined by the slatted-metal fence of the lot itself, a third side is made of wood, a fourth is open but equipped with screens that can be hung when necessary to block wind or sunlight. Fresh garbage is stored in screened bins and in the kind of large freezer in which you find bags of ice at convenience stores; the freezer was brought in after members of a Garbage Project expedition to sort garbage in Milwaukee in the middle of winter noticed that frozen garbage didn't smell very much and didn't attract flies or hatch maggots. ("The freezer," says Wilson Hughes, "has done for garbage studies what the microchip did for electronics.") Garbage sorters learn to sort quickly so as to finish before the garbage has thawed.

The Garbage Project collects its garbage in different ways for different studies, but in essence the methods fall into two broad categories. The first, known as a regular sort, involves dividing neighborhoods into groups based on income levels, family size, and educational attainment (as identified by the U.S. Bureau of the Census); in each of these neighborhoods sanitation workers then collect for the Garbage Project all the garbage placed out for city pickup by randomly selected households. The duration of the periods during which refuse pickups are collected varies, depending on the study, but Garbage Project collections for some purpose are conducted from February through May and September through November, and sometimes through the summer as well. Garbage collected in this way can be used either to compare the discard patterns of different types of neighborhoods or to obtain snapshots over time of discard patterns in the aggregate.

This regular sort is a straightforward procedure, and immensely valuable. It has yielded, for example, most of the overall data on how much of every type of food ends up as waste. The figure for the edible part of bananas, to give one example, is about 8.5 percent; for baby food, about 10 percent; for breads and cereals, as much as 15 percent overall; the figure for the much-maligned prepared-food category (which includes microwave dinners, take-out food, and packaged soups and stews) is a relatively modest 4 to 5 percent. The food category that undoubtedly has far and away the most positive public profile—fresh produce—is also far and away the biggest contributor to food waste: Produce accounts for from 35 to 40 percent of total edible-food discards by weight. This figure does *not* include thrown-away portions of produce that aren't really waste—rinds, peels, skins, and so on—which constitute a considerable category unto themselves. By weight, the inedible part of an avocado is some 24 percent of the total; of a banana, 32 percent; of a lemon or grapefruit, 50 percent. In terms of garbage generation the lowly potato peel is a powerhouse among comestibles: Of all food that is thrown away (not just "waste" but also inedible rinds and tops), potato peels account for a mighty 7 percent by weight—the largest single item in the fresh-food repertoire. Potato peels are so prevalent that they have earned the honor of a distinct Garbage Project code number (044). In terms of overall weight, edible and inedible food debris accounts for about a fifth of household garbage—a potentially substantial source of compost.

The regular sort of garbage can also point up certain broad phenomena—as, for example, that after Halloween one finds lots of candy wrappers and almost no candy in garbage, while after Valentine's Day one finds that the candy itself (along with the wrappers, which still enclose it) often gets thrown away. One Garbage Project researcher, Jeffrey Parks, used the data from several thousand random pickups to investigate whether, owing to the growing prevalence of AIDS and the well-publicized admonitions to practice safe sex, the use of condoms had perceptibly increased in homes. He found that the number of condom wrappers identified in garbage per every one hundred garbage pickups remained stable between 1976 and 1984 but increased by 45 percent between 1985 and 1987. (Condom wrappers are used by the Garbage Project as a "marker" for the condoms themselves, because condoms are more often flushed down the toilet than thrown away in a trash can; sometimes God has mercy on garbologists.) Investigations into the use of contraceptives by women have brought disconcerting practices to light. For instance, an analysis of birth control pill dispensers found in garbage showed that a good many of the women using the pills seemed to have been using them incorrectly. One of the monthly dispensers showed a single pill missing, and it had been removed from the very middle of one row of pills. From another dispenser, every second pill had been taken.

Additional studies using Regular Sort methodology have focused on alcohol. Does the opening of a liquor store in a neighborhood that didn't previously have one effectively cause the people in that neighborhood to drink more than they used to? There is some evidence that it does, judging from a Garbage Project study by a University of Arizona undergraduate, Shannon McPherson, of the discard of bottles and cans in various neighborhoods before and after the opening of new liquor stores. Do people drink more when the moon is full, as popular folklore would have it? A Garbage Project study by Frederic Haskell, another undergraduate, of the number of beer cans and bottles discarded during the various lunar phases over a period of twelve years revealed no correlation between heavy beer drinking and *any* phase of the moon; there was, however, a correlation between heavy beer drinking and paydays.

What about the question: Do the poor pay more? That is, do poor people, perhaps strapped for cash, buy smaller amounts of products each time they shop than more-affluent people do, thereby missing the savings that buying in bulk makes possible? Economists have long suspected that this is the case; the hypothesis is supported by Garbage Project research based on a comparative study of differing neighborhoods in both Tucson and Milwaukee. Lower-income families consistently buy small-sized packages of everything from cereal to detergent; more-affluent families consistently buy the "giant, economy size." No maker of house-brand or generic laundry detergent makes detergent available in a small, twenty-ounce container;

some of the makers of brand-name detergents do. These brand-name, twenty-ounce detergent containers are, if found in garbage at rates well above average, a telltale sign of a low-income neighborhood. The same phenomenon is visible in discard rates of large and small cans of solid food. A Garbage Project study of large families in two relatively rich and two relatively poor census tracts in Tucson during the worsening "stagflation" of the mid-1970s showed clearly that the affluent responded to hard times by increasing the amount of canned food they bought in large (bigger than sixteen-ounce) cans, while during the same period the proportion of large cans in the garbage of the poor declined by almost 50 percent, and the proportion of smaller cans rose. Such patterns of consumer behavior, caused by economic deprivation and not personal perversity, point up one more way, albeit a subtle one, in which the bonds of poverty are secured. They also point up a terrible irony with respect to the generation of garbage, at least insofar as supermarket buying is concerned: Because the ratio of product to packaging is so much higher among purchases by the affluent than among those by the poor, the poor end up throwing away more packaging per ounce of useful product than the affluent do. This fact could have implications for humane planning. From time to time there is serious talk about the possibility of adding a "product-disposal charge" to the cost of packaged consumer items to help defray the cost of their disposal. Whether such charges would be a good idea remains an open question, but a consideration of their disproportionate impact on the poor ought to be taken into account.

The second methodology employed by the Garbage Project combines the garbage sorts of the first methodology with an additional element: verbal "self-reports" about personal behavior, elicited by interviews or recorded in daily diaries, from the very people whose garbage is being collected and sorted. In this way not only can discard patterns be matched one-to-one with a household's socioeconomic characteristics—income, family size, and so on—they can also be matched against what the people who have been doing the throwing away think or say they have been throwing away. Conducting what is known as a "matched study," of course, means that the Project must secure the active cooperation of all the households involved for a period of as long as five weeks—not an easy proposition. In the Garbage Project's two-decade history of matched studies only a few of the people asked to participate have objected to having their garbage collected and sorted (one person who did object was a new mother of twins, who worried about the sorters having to deal with all those diapers; she relented when assured that the sorters had seen worse), but as many as four people out of five, perhaps not surprisingly, refuse to participate in the essential second part of the research effort, the personal interview. As noted earlier, the anonymity of participants in matched studies is guaranteed. The garbage records and interview data from any given household are correlated one to the other only by number codes. Names and addresses are not retained by the Garbage Project.*

Studies that use this second methodology—studies that compare a household's perceptions of its garbage with the actual garbage that the household generates—always reveal some telling human quirks. Foremost among them, as one might suspect, is the tendency of people to be unreliable sources of quantitative information about their behavior. What people claim in interviews to have bought and consumed, to have eaten and drunk, to have recycled and thrown away, almost never corresponds directly or even very closely to the actual remnants of material culture in their Glad or Hefty bags (see Figure 3-B).

Consider the matter of food waste. In 1980 and 1981 the Garbage Project conducted an intensive study of food waste in sixty-three households. The study was done on behalf of the U.S. Department of Agriculture, which for obvious reasons has long had an interest in the ultimate disposition of the nation's food. What the USDA did not have was any notion of the accuracy of the information it had gathered on food waste by means of its Nationwide Food Consump-

* They are, however, retained in a file drawer of the anthropology department at the behest of the University of Arizona's Human Subjects Committee, which, like its counterparts at most universities, was set up to protect the rights of people who become the object of social-science research. The file drawer holds forms signed by participants in matched studies, bizarrely acknowledging their understanding that their anonymity is assured. There is still no way, however, to associate specific individuals with the garbage for which they are responsible.

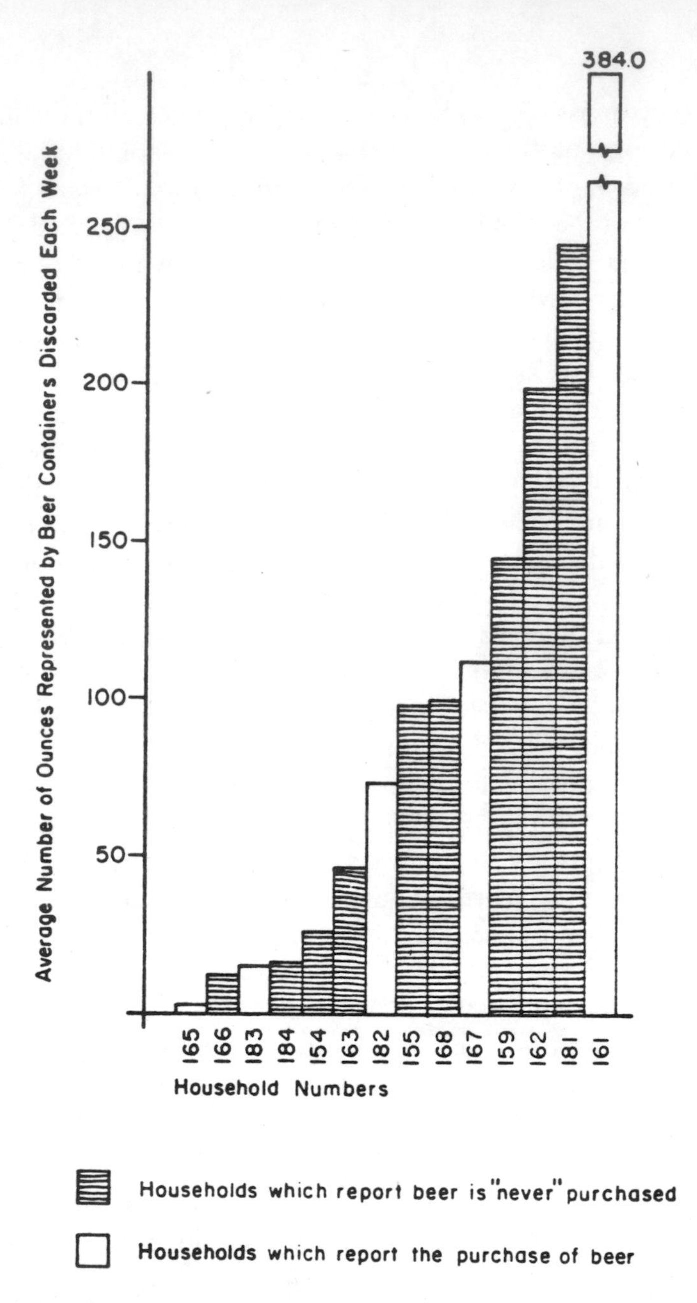

Figure **3-B.** Garbage Project research frequently points up the unreliability of certain kinds of information acquired by means of surveys. A case in point involves the results of interviews with members of many of the households represented above, households which have in common the fact that the garbage from every one of them was found to contain beer bottles or beer cans. The data here were gathered over a five-week period and are presented as a weekly average.

SOURCE: The Garbage Project

tion Survey—that is, by asking people what they could tell interviewers about food waste in their own homes. For its own food-waste study the Garbage Project collected and sorted garbage for five weeks from all of the sixty-three sample households, which were each identified by a code number. For the purposes of a self-report, however, the households were divided into four groups. All of the respondents in these households were carefully briefed as to what was to be considered "food waste" (as opposed to unavoidable "food preparation debris," such as peels and rinds, bones and skin), according to the Garbage Project's definition. And all of the respondents knew that university students would be combing through their garbage and keeping precise records of what was found.

None of the households in any of the four groups reported anything like the actual amount of edible or once-edible food that the sorters pulled from their garbage. The members of group one delivered their self-reports through interviews, in which they were simply asked to recall how much food they had wasted in the previous week. No one reported much wasted food at all, although Garbage Project sorters in fact found in group one's garbage an average of two-and-a-half ounces (more than an eighth of a pound) of edible or once-edible food per household member per day.

The members of group two were asked to keep written records for one week of the amount of food they wasted (in the manner of households that keep diaries of television viewing for the A. C. Nielsen Company). All of the households agreed to comply, but at the end of the collection period group two logged only a handful of reports of wasted food, and the food in these instances was described as mere scraps. Meanwhile, Garbage Project sorters were likewise turning up more than an eighth of a pound of wasted food per person per day in group two's garbage.

The members of group three were given scales and asked to weigh any food they wasted over a one-week period of time, and to keep records of the weights day after day. You will surely have guessed how much food group three reported wasting—not very much. The actual amount was again a little more than an eighth of a pound per person a day.

Group four proved to be something of a surprise. The households in group four were provided with boxes of plastic bags and asked to use the bags over a period of three days to store any food that would normally have been thrown away. Garbage Project personnel visited the group-four households every night to collect the bagged food waste. After three days the amount of food waste collected in the plastic bags turned out to be quite a lot: about one-quarter of a pound per person per day. And yet the sorters picking through group four's garbage cans still found a lot of food waste—an eighth of a pound per person per day, just as they had found in the garbage of groups one, two, and three. The respondents did aim to please. When the Project seemed to be checking up on people's waste habits, the self-reports were tailored to reveal households in the best possible light. When the Project, as in the case of group four, seemed to want wasted food, it got wasted food. Meanwhile, everyday food-waste behavior continued unabated.

The Garbage Project has, over the years, conducted a variety of analyses of matched data for the Department of Agriculture, all related to the evaluation of the Nationwide Food Consumption Survey; among the results has been the identification of several behavioral syndromes that seem to occur repeatedly.

One of these is the Good Provider Syndrome: Almost uniformly, homemakers report that their families consume considerably more food than sorters can actually find evidence for in the household's garbage. They also report consuming much more food than is indicated by the sum of the personal self-reports provided by the individual members of the household. Presumably, the homemaker is overreporting—no doubt unconsciously—in order to demonstrate that the household is amply supplied with life's necessities. (Perhaps for a similar reason, homemakers tend to underreport the amount of prepared foods the family uses, and to overreport the amount of fresh produce employed in cooking meals from scratch.) It is also the case that individuals in their self-reports are minimizing the volume of certain kinds of food intake—a phenomenon that might be called the Lean Cuisine Syndrome. People consistently underreport the amount of regular soda, pastries, chocolate, and fats that they consume; they consistently overreport the amount of fruits and diet soda. The following tabulation, compiled by Susan Dobyns, who is the Garbage Project's research director, is taken from one of the Project's USDA studies, and shows the amounts of various food items that were overreported and underreported. It is a sad catalogue of self-delusion:

	% Underreported		% Overreported
Sugar	94	Cottage cheese	311
Chips/popcorn	81	Liver	200
Candy	80	Tuna	184
Bacon	80	Vegetable soup	94
Ice cream	63	Corn bread	72
Ham/lunch meats	57	Skim milk	57
Sausage	56	High-fiber cereal	55

A phenomenon related to the Lean Cuisine Syndrome is the Surrogate Syndrome: People may provide inaccurate consumption reports about themselves, but if you ask them to describe the behavior of a family member or even a neighbor, they tend to squeal with chilling accuracy—especially when the behavior involved has a negative image. With respect to alcohol intake, for example, most people underreport their drinking by 40 to 60 percent; a surrogate in the same household who does not drink alcoholic beverages and who is asked to report on the habits of members of his family will get intake levels right to an accuracy of about 10 percent.

In sum, the data generated by the Garbage Project in its cross-check of the USDA's Nationwide Food Consumption Survey reveal that much of the information in the government's vaults about food consumption and waste may be shaky stuff indeed. But perhaps reassuringly, the data also indicate that a great deal of what we have all suspected about certain tendencies in human nature appears to have been suspected with no little justification. As T. S. Eliot once observed, "Human kind cannot bear very much reality."

Garbage Project behavioral studies always seem to come back to the subject of meat, no doubt because meat is at once a key element in the American diet, a primordial element in psychological perceptions of well-being, and a comestible about which many people harbor feelings of deep ambivalence. In addition, as noted earlier, the nature of its packaging is, for Garbage Project purposes, at once valuable and compelling. Of course, the remains of meat in fresh garbage or at a landfill are not the most pleasing archaeological artifacts that one can handle. The packaging is bathed in blood and in a kind of clear slime whose viscosity feels to the touch like that of molten gristle. The associated artifacts include the familiar absorbent deli pad, or "meat diaper," that drips sanguinary rivulets on the garbage-sorters' aprons. This is the kind of experience that sorting entails, and there's no getting around it. In the sorting yard, as the garbage is dumped onto tables, one person in thick gloves picks through it, looking for all the world like a surgeon with his hands deep in a patient's entrails, and calls out information to a second person, standing some distance away, who carefully records it (see Figure 3-C). Once entered into a computer's memory, the data are available for analysis.

Meat-wrapper labels provide a demonstration in microcosm of the tenuous grasp many of us have on even the most familiar of objects and behaviors. How are meat weights recorded on a package? In pounds and ounces, most people will say. In fact, they are written in pounds and tenths of pounds and hundredths of pounds. Two market researchers, Helen C. Brittin and Dale W. Zinn, found in the course of a 1977 survey that 40 percent of all shoppers interviewed as they left a supermarket misreported the amount of meat they had just bought by significant amounts. The Garbage Project has found a similar pattern in comparing what people say they have prepared and eaten in the way of meat with what is indicated by discarded packages, bones, and cut-away fat. Some people simply can't remember cuts or quantities; others either consciously or unconsciously misreport.

One of the most consistent patterns with respect to meat is what can be called the Spike Effect. When asked about quantities of meat bought or prepared, respondents tend to round off the numbers into pound (sixteen ounce) or half-pound (eight ounce) increments—who knows what .17 pounds really means? The pervasiveness of this tendency is striking; for example, when asked about their consumption over a certain period of time, in ounces, of such things as red meat, poultry, cheese, saturated fats, and pork bacon, in 80 percent of all cases respondents gave figures that were divisible by eight,

8.32 TOTAL HH WEIGHT (LBS.)

CENSUS TRACT (16 17 18)	COLLECTION MO. (19 20)	COLLECTION DAY (21 22)
A46	02	08

NAME OF RECORDER:

DATE OF ANALYSIS: 2/9/91

MATERIAL COMPOSITION CODES

CODE (LIST MOST PREVALENT MATERIAL FIRST)

A PAPER	H RETURNABLE GLASS	R COPPER AND BRASS
B FERROUS (STEEL/TIN)	J AEROSOL CANS	S BIODEGRADABLE PLASTIC
C ALUMINUM	K WOOD	T TEXTILES
D PLASTIC (CELLOPHANE)	M CERAMICS	V CORRUGATED CARDBOARD
E NON-RETURN GLASS	P LEATHER	X OTHER (SPECIFY ON BACK)
	O RUBBER	

TRACT	PAGE / TOTAL PGS.
A46	1/2

	ITEM CODE (23 24 25)	NO. OF ITEMS (26 27 28)	FLUID OUNCES (29–33)	SOLID OUNCES (34–38)	COST (39–44)	WASTE (GRAMS) (45–48)	SPECIAL IND. (49)	BRAND (50–57)	TYPE (58–65)	MATERIAL COMPOSITION CODE (66 67 68)
1	011	2	256.	.	.			JACKSONS	2PERCENT	D
2	024	1	.	14.	.			UNCLEBEN	INSTANT	A
3	098	1	.	15.	.			HORMEL	BEEFHASH	B
4	097	1	5.	.	.			CHUNKING	SOYSAUCE	ED
5	095	1	8.	.	.			MCDONALD	COFFECUP	D
6	002	1	.	17.9	3.35			FRYS	PORKCHOP	AD
7	140	1	.	.	.			WALMART	DEPARTME	A
8	140	1	.	.	.				FREEZER	D
9	006	1	.	16.6	7.27				SHRIMP	A
10	027	1	.	8.	.			PILLSBUR	FROZEN	AB
11	134	1	.	.	.			MRCOFFEE	FILTERS	A
12	070	1	.	.	.				GROUNDS	
13	079	1	67.6	.	.			SQUIRT	LEMONLIM	D
14	055	3	.	.	.			WRIGLEYS	GUM	
15	111	14	.	.	.			LUVS	DIAPER	TA
16	048	30	.	.	.	520			GRAPES	
17	001	1	.	16.	1.61			SMITHS	GROUND	AD
18	045	1	.	48.	.			STRAND	APPLE	D
19	073	2	.	.	.			LIPTON	BAG	A
20	136	1	.	.	.				ROLL	A

Figure 3-C. A typical recording form. In the course of the Garbage Project's history, forms like these have been filled out by upwards of 500 student garbage sorters. The Garbage Project's database includes information from some 55,000 of these documents.

SOURCE: The Garbage Project

which of course is half the number of ounces in a pound. (Oddly enough, researchers run into a similar phenomenon, known as "age-heaping," when they look at large numbers of self-reports, as rendered on Census Bureau forms, of people's ages. In this case the spikes come at ages divisible by five, and reflect rounding off.)

What is more, the direction of the rounding, up or down, is not random. In upper-income neighborhoods and upscale retirement communities—places where concern for healthful living was evident from the number of health-related magazines in garbage and other suggestive indicators (for example, the relatively large proportion of shelf space in local supermarkets reserved for health-related products, such as high-fiber cereals, and the presence in supermarkets of automated blood-pressure gauges)—people tended to round down the amount of beef and pork they reported eating. In lower-income and middle-income neighborhoods, where the eating of meat is often viewed as an indicator of status, people tended to round up.

One important Garbage Project study involving red meat was focused on the question of fat consumption over time. Meat fat has always been a problem for human beings. For hunters like those who butchered the bison at the Olsen-Chubbuck site 8,500 years ago, the problem was not too much fat but getting enough fat in their diet during the winter, when animals (and, consequently, the people who ate them) grew lean. In recent times, of course, the problem has been the opposite: a surfeit of fat in our diet all year long. Fifty years ago people often ate their beef and pork cuts fat and all, and many nutritionists have wondered about the extent to which Americans today may have altered their habits in this regard. The Garbage Project's refuse sorts offered one way to investigate the issue: There is discarded fat to weigh, there are wrapper labels to indicate type of cut and original weight, and there are USDA conversion tables to determine the amount of separable fat that a cut of meat initially had. With all this information it was possible to get some idea of basic trends over a period of years in the trimming and discarding of meat fat. (Bear in mind, however, that it is not possible to determine absolutely what percentage of all fat is not being eaten by people; dogs, cats, and garbage disposers gobble up much of the pertinent data before sorters can get their hands on it.)

What the investigation revealed was that between 1976 and 1982, the percentage of fat trimmed off meat and then discarded held stable. Beginning in 1983, the percentage of fat trimmed off and discarded suddenly doubled in all of the Garbage Project's sample neighborhoods, including a high-income community in the San Francisco Bay area and a retirement community south of Tucson, and meat-fat discards have remained at this elevated level ever since. Why the greater discard of fat? There had been no diminution of the dog and cat population during the course of the study, and the proportion of homes with garbage disposers in the sample neighborhoods had actually increased. The only variable that could be found to explain this widespread shift in behavior was the publication, at the end of 1982, of the National Academy of Science's report *Diet, Nutrition, and Cancer,* and the subsequent onslaught of reports in the media identifying fat in the diet, particularly fat from red meat, as a significant cancer risk factor.

People, it would seem, were beginning to behave a bit more sensibly. And yet, as a closer look soon made clear, people are incorrigible. For even as Americans have been separating more fat from their fresh meat (and also, it was discovered, buying less fresh meat overall), they have been eating a higher percentage of red meat in processed forms (hot dogs, bologna, salami, bacon, sausage), which contain large quantities of hidden fat—far more, for example, than in the marbling of a steak. In other words, while people are buying fewer fresh cuts of beef, and trimming more fat off what they do buy, they are compensating for their good behavior—no doubt unwittingly—by ingesting meat fat on the sly. Among middle-income households, it may be that the hidden fat in convenience meats and processed meats now accounts for 70 percent of all meat fat consumed. Plainly, people have somehow grown accustomed to thinking of "fat from red meat" as synonymous with "separable fat from *fresh* red meat"—in the process overlooking a second (and ubiquitous) category of meat fat.

Hardly a Garbage Project study has been conducted that does not depict the average American as fundamentally unaware of some of the most familiar activities in which he or she indulges—and unaware, too, of how odd, even disturbing, are the consequences of some of those activities. One example involves the disposal of hazardous household waste. In 1986, 1987, and 1988, Garbage Project sorts were conducted in New Orleans, Phoenix, Tucson, and Marin County, California, in order to identify the types (for example, oil-based paints and stains, photographic chemicals, oven cleaners) and quantities of hazardous waste that are discarded in everyday household trash. It turns out that about 1 percent of all household garbage by weight consists of hazardous waste—not very much, one might think, until one remembers that this 1 percent, when multiplied by, for example, the 88,000 households in Marin County, yields an annual total of 64,700 pounds of toxic chemicals going into Marin County's two landfills.

While the proportion of hazardous waste in household garbage does not seem to vary among neighborhoods with sharply different socioeconomic characteristics, the composition of the hazardous waste was found to vary considerably. The hazardous waste from low-income households consisted disproportionately of car-care items: motor oil and gas additives especially. Middle-income households, in contrast, seem to lavish less attention on their cars and more on their homes; their hazardous waste consisted disproportionately of paints, stains, varnishes, and various other products associated with a dedication to home improvement. The garbage placed out by households in affluent neighborhoods reflected the greater attention paid there to lawns and gardens: It contained unusually high amounts of pesticides, herbicides, and fertilizers.

Needless to say, homeowners, when interviewed, had little idea of what kinds and quantities of hazardous waste they were throwing away (if they were aware that they were throwing away any such waste at all). In affluent Marin County, homeowners reported that car-care products probably accounted for most of the hazardous waste they generated; they rarely mentioned lawn-care products; in fact, their garbage looked a little like the remains of an agribusiness yard sale. It should be noted, by the way, that the First Principle of Food Waste seems to have an analogue, the First Principle of Household Hazardous Waste. As the Garbage Project's Douglas Wilson has determined, products such as cleansers and detergents that are used on a regular basis exhibit very little waste; products such as sealants and glues that may have been bought for a single specific renovation job around the house account for an immensely disproportionate share of household hazardous waste.

In the course of conducting these toxic-waste studies Garbage Project researchers noticed an unwelcome phenomenon. In many places across the country, including Marin County, local communities sponsor special collection days when residents can bring any household hazardous waste that they want to get rid of to a centralized location, from which the waste will be transported either to a recycling facility or to an official Subtitle C hazardous-waste-management facility. In 1986, in order to determine how effective these special collection days actually were in reducing hazardous waste in household refuse, the Garbage Project sorted samples of Marin County garbage one month before and two months after the county's first well-publicized "Toxics Away!" Day. The results were contrary to what had been expected. The garbage discarded *after* "Toxics Away!" Day contained more than twice as much hazardous material by weight as the garbage that had been discarded *before* "Toxics Away!" Day. Somehow, "Toxics Away!" Day, which was intended to channel Marin County's household hazardous waste into the hands of people who would dispose of it safely and professionally, also had the effect of increasing the amount of such waste that was simply being sent to the county landfill.

Why was this the case? The culprit was probably the intense media campaign designed to make county residents aware of the collection day. Because the collection was held on only one day, and no future collection days were announced, it seems likely that many homeowners who had been made newly aware of the hazardous waste in their homes but had missed the collection day decided to rid themselves of their hazardous waste in the conventional manner: They just threw it out. The "Toxics Away!" Day phenomenon has since been confirmed by studies in both Tucson and Phoenix. The solution to this problem would seem to be either to have more-frequent collection days or a permanent collection facility.

As the foregoing suggests, it seems at times as if garbage is capable of wielding some sort of malign influence: that its very presence acts to bend rational minds to irrational ends, and to thwart the noblest of efforts. But, to be fair, garbage on occasion has also figured in happier (if minor) behavioral episodes. Michael Owen Jones, a specialist in folklore and folk art at the University of California at Los Angeles, provided an example in the course of a slide presentation he once gave at an academic conference, in which he sought to demonstrate how patterns in certain mundane daily communal activities could in themselves constitute a kind of folk tradition. Jones's attention (and camera) had been drawn to the curious ways in which his neighbors in a Los Angeles suburb put out their trash for collection. Apparently the residents of the neighborhood, where garbage had to be left out for collection in front by the road (there were no alleys), visible to all, had been gradually overcome by a sense of public shame, or perhaps a mild dementia. First one homeowner, then another, then a few others began spending considerable time and energy arranging their garbage in pleasing ways:

> My neighbor to the north, for instance, [slide], has a matched set of plastic containers. He places them in a perfect line along the edge of the curb, lids in place to contain the trash neatly, and the handles exactly aligned. When on one occasion this fall, after having raked fallen pine needles from his yard, he had too much trash for his set of containers [slide], he lined up his cans as usual and behind them—using a single type of container of the same size and shape and color, just like his plastic cans—arranged paper bags filled with needles. A neighbor across the street usually lines his cans with plastic leaf bags, and arranges the four cans in a square. Another homeowner, whose driveway is bordered by a low wall, and whose mailbox is near the entrance to the drive [slide], places his trash cans to either side of the mailbox in a visually balanced arrangement.

And as Jones himself went on to report, even he, the cooly distant observer, was ultimately unable to resist the force that held the neighborhood in its thrall: For there came a day, he admitted, when he replaced his delapidated mongrel garbage cans [slide] with a handsome matched pair. The Joneses had kept up.

PART II

The Landfill Excavations

CHAPTER 4

INTO THE UNKNOWN

Anyone who has made the trip along the upper gullet of the New Jersey Turnpike as it disgorges traffic toward the George Washington Bridge is familiar with the Hackensack Meadowlands. This vast glacial fen runs roughly from Newark, New Jersey, to Nyack, New York, parallel to the Hudson River, and is separated from Manhattan by the Hudson and the long, craggy spine of the Palisades. The southern approach to the Meadowlands is heralded by a grim landscape of chemical plants and refineries, licks of flame dancing beneath the hazy gray of the sky. Across the Meadowlands are twisted roadbeds at what seem to be arbitrary and unnecessary heights. Below, small pools and channels can be glimpsed among the tall reeds, and there are moments even now, as a turn in the road affords a certain view, when the awesome sweep of this wetland in its nativity can be imagined still.

Such moments, of course, are rare. More commonly the eyes take in the massive mounds of garbage, some of them fifteen stories high, that have been dumped in the Meadowlands—blanketed in some cases by a film of dirt, and picked over every second of the day by a scavenging of gulls. More than a hundred communities once dumped their garbage into the Meadowlands, and garbage dumps cover three square miles of it. Almost all of this garbage was deposited in the days before measures were routinely taken to prevent or minimize seepage (as is now mandated by federal regulations). While systems to vent methane gas and control leachate exist at a few of the Meadowlands repositories, most of whatever is leaking out of the vast majority of them is leaking right into the water—into the Hackensack River, eventually, and then into New York Harbor. All but one of the Meadowlands dumps are now shut down (the last covers a mere seventeen acres) but the damage has been done. The mounds may not be permanent eyesores—skillful landscaping has beautified many such sites—but their contents could foul the area for decades to come.

The Australian archaeologist Rowland Fletcher calls the largest monuments that any society builds for itself MVSes—Monstrous Visual Symbols. Fletcher has noted that over the centuries, as a society's motivating ideals undergo change, so do its MVSes: from, say, temples and cathedrals to bridges and skyscrapers. The Hackensack Meadowlands are a potent reminder that the largest MVSes in American society today are its garbage repositories. Archaeologists believe that the biggest prehistoric MVS in the New World is the Pyramid of the Sun, at Teotihuacan, which was built in Mexico around the time of the birth of Jesus. Its volume is 75 million cubic feet. The garbage dumps in the Meadowlands exceed that volume many times over, as do most big-city landfills. In the San Francisco Bay area, the volume of the Durham Road landfill has already reached 150 million cubic feet; it has been built from the municipal solid waste of three moderate-sized towns over a period of only fifteen years. Fresh Kills, of course, is many times larger still. These MVSes may not be Chartres, but they are not without a certain grandeur. Many are surrounded by low brush which snags the thousands of thin plastic bags of various hues that blow from the dumping site, and at dawn the sun lights this perimeter in vibrant color.

Landfills are fitting symbols of many of the developed world's twentieth-century preoccupations—and they are great wellsprings of mythology as well. It is somehow fitting that the Hackensack Meadowlands Development Commission chose in 1989 to lodge a garbage museum in the environmental center at DeKorte State Park, which covers a two-thousand-acre tract of not-quite-pristine wetlands that abuts a ridge of dumps. One striking floor-to-ceiling exhibit through which visitors are able to walk is a bright, cavernous jumble of trash. The structure is the work not of a sanitation professional but of a thirty-year-old artist from Newark, Robert Richardson, whose intentions included making visitors feel that garbage was about to engulf American society. "They'll feel that the garbage climbing up the walls is overwhelming and at some point might fall over," Richardson told a reporter. "That's good."

To most visitors the contents of the display no doubt seem visually synonymous with the contents of American garbage in general, and thus with the contents of a typical landfill. Look: There are the empty boxes of Brillo and Tide, the plastic jugs and protective foam cartons, the disposable diapers, the bottles and cans, the fast-food packages—all of these things, assuredly, items that do get thrown away, that one does find in garbage and in landfills. But the popular perception of garbage sometimes does not accord fully with reality. If a worker from the local department of sanitation were invited over to the garbage museum at DeKorte State Park and asked to point out to visitors how the garbage he has to deal with every day differs from the garbage displayed in Robert Richardson's construction, he might note, to begin with, that there seems to be no dirt mixed in with this garbage, and yet each day's deposits in a real landfill are tucked in with a layer of dirt. He might note that there is no construction and demolition debris, and no food and yard waste or, indeed, organic waste of any kind—no grease-soaked newspapers, no discarded trays of kitty litter, no sewage sludge. (He would, of course, understand *why* there was no organic waste at the museum; it is for the same reason that verisimilitude is kept at bay in colonial Williamsburg.) Our visiting sanitation worker might note that there is a good deal more plastic on display at the garbage museum than you would actually find in most landfills, and a lot less paper. He might note that none of the garbage appears to have been crushed, even though most garbage in a real landfill looks as if it has been run over by a forty-two-ton compactor, which it often has. And he might conclude with the obvious observation that the garbage on display gives off no smell—perhaps venturing to remark, and speaking as a connoisseur, that the bouquet of a well-managed sanitary landfill, though it hangs more thickly than more desirable atmospheres, is not entirely unpleasant.

How wide the gap may be between garbage myth and garbage reality surely varies from one specific issue to another, but there is probably no issue relating to garbage where a gap does not exist. In the Meadowlands garbage museum a life-sized, three-dimensional tableau depicts a twentieth-century American family blithely throwing away plastic cups and sheets of aluminum foil; instead of faces, the display's human figures have mirrors, inviting visitors to see themselves in similar situations. Those mirrors are apt symbols of much of the conventional wisdom about garbage, which often simply reflects the misinformation that people bring to the subject. The result, inevitably, is a closed system of fantasy and shortsightedness that both hampers the effective disposal of garbage and leads to exaggerated fears of a garbage crisis. A growing body of research findings from Garbage Project landfill digs and other investigations has begun to provide redress. We will look at that research in a moment, after a brief excursus into the history and architecture of the sanitary landfill.

From the perspective of history, the idea that modern landfills should now be deemed to be a major social problem—which is certainly the widely accepted view—is rather ironic. The sanitary landfill started out in life as a solution to the twin problem of garbage incinerators that befouled the air and the malodorous open dumps that ringed American cities like vile garlands. In the United States there still exist a multitude of open dumps, a few of them official or semiofficial repositories, many more of them representing informal and illegal accretions of garbage. As Garbage Project and other studies (notably John Hohmann's "Trail's End" study mentioned in chapter three) have pointed out, any deserted area where a road suddenly terminates is likely to serve as a local dumping site. The pattern is all too familiar. The immediate roadside area is littered with odds and ends. In a broader circle beyond are beer and soda cans and broken bottles —probably tossed from cars. Beyond them are thrown-away durable goods: abandoned cars, decomposing sofas, rusting mattress springs.

These road-end sites are eyesores at the very least. Those (mostly in rural areas) where organic garbage is still discarded can be disturbing structures indeed, to both eye and nose.

Taken as a whole, illegal dumping sites of one kind or another are surprisingly numerous; one Garbage Project survey by a University of Arizona undergraduate, Steven Clifford, of the accessible desert around Tucson found a total of more than seventeen hundred such sites of various sizes, most consisting of what appeared to be one big load of one household's garbage. But while these dumps may be unsightly and numerous, the percentage of any city's garbage that is disposed of in open dumps is quite small. Most of the garbage now goes instead to that enduring legacy of the Progressive Era, the sanitary landfill, a repository whose operations are today regulated by an increasingly stringent but by no means perfect web of state and federal strictures. Owing to budget constraints and lax enforcement, about half of all sanitary landfills in operation today are operating without permits.

A sanitary landfill, in its simplest form, is one where every day all the new garbage that has been hauled in is covered with six inches or so of some material that is relatively inert and won't decompose: soil, mainly, although crushed glass and even a plastic foam (one brand is called "Sanifoam") have been used. The civil engineer Charles Gunnerson made note, in his study of refuse accumulation in ancient Troy, of the parallel reliance in that city and in cities of our own time on covering garbage with layers of dirt; he found modern landfill management "reassuring" as a result, and indicative of "the role of the earth in assimilating wastes and controlling odors since ancient times." The dirt cover also helps to keep pests to a minimum.

Who invented the modern landfill? Most conventional accounts say that the British did, in the 1920s; the procedure was known in Britain as "controlled tipping." But according to the historian Martin Melosi, one can find examples of something like sanitary landfills in America even earlier: in Champaign, Illinois, in 1904; in Dayton, Ohio, in 1906; in Davenport, Iowa, in 1916. Wherever the concept first happened to appear, the impetus was a concern for public health. Even before the role of bacteria and viruses in the onset and spread of disease was well understood, people had made the connection between sickness in the community at large and the open dumps nearby. This perception inhered in the now discredited "miasmic theory" of disease, which attributed contagion to poisonous gases that were said to emanate from sewage and rotting organic debris. The specter of "miasmas," invoked repeatedly by public-health officials and newspaper editorialists, spurred urban cleanup efforts on a broad front. As is frequently the case in the history of human progress, some good things ended up happening for all the wrong reasons.

The sanitary-landfill idea at first caught on very gradually, though by the 1930s a number of examples could be found on both coasts, in New York and California. The term "sanitary landfill" itself seems to have been coined in the early 1930s by Jean Vincenz, the commissioner of public works of Fresno, California. The procedure received perhaps its biggest boost during the Second World War, when the Army Corps of Engineers adopted it as the disposal method of choice for U.S. military facilities—a move that had the twin effects of making millions of servicemen aware of sanitary landfills and training thousands of people to operate them. Sanitary landfills came to be regarded as an obviously preferable solution to smoke-belching incinerators—the acrid means used by most cities at mid-century for getting rid of the bulk of whatever garbage they managed to collect. And landfills, which were designed to be covered over and landscaped or even built upon once their life as landfills was at an end, could, if placed in the right locations to begin with, help turn marginal terrain, such as wetlands, into productive real estate. The notion that one could both solve a problem and in so doing create wealth has always held powerful appeal for Americans. It is hardly surprising, then, that possession of a landfill was seen as a hallmark of a well-managed city. By 1945, about a hundred American cities had created sanitary landfills. Within fifteen years the figure was fourteen hundred.

The exact recipe for the perfect landfill has changed with time and the popularity of various theories. Many of the basic principles, however, have remained constant. The first consideration—and the one for which the criteria have changed most radically—is the site itself. As noted, it was once believed that sanitary landfills should be situated in such a way as to help reclaim wetlands and other low-lying areas. This view has turned out to be doubly wrong: Wrong because the environmental importance of wetlands was not well understood, and wrong because the hazards of the liquids that may drain out of landfills were also not well understood. As a result, many of the earliest landfills were put in the worst places imaginable, and we are living with the consequences. Much of the animus directed at new landfills has its origin in the nasty reputation of old ones.

Today the emphasis is on using hydrogeologic studies to site landfills in places where contamination of ground and surface water can be avoided. Rainfall runoff patterns are taken into account, and sites are chosen, ideally, where the underlying matrix—that is, the *configuration* of underlying soil and rock—has a hydraulic conductivity in the range of 10^{-6} or 10^{-7} centimeters per second. The 10^{-6} range would include matrices that consist of silt-clay-and-sand mixtures, silt-and-clay mixtures, laminated sandstone, shale, and mudstone. The 10^{-7} range would include matrices that consist of a formation known as "massive clay," and also large formations of igneous and metamorphic rocks. The point is: The best sites are those where fluids will have considerable difficulty making their way beyond the landfill's boundaries and into bodies of water. Some places in America are so geologically unsuited for landfills—most of Long Island and much of Florida, for example—that building new ones there is virtually out of the question.

At some landfill sites garbage is simply piled on the earth's unbroken surface, but more often the next step is to dig a great hole—one that is usually from twenty-five to fifty feet deep, though it can be deeper. On occasion, a cavity may already exist at an appropriate site: A large number of holes have been dug in the United States—and never refilled—in the course of extracting coal, copper, gravel, and other natural resources. However, these holes are almost always either too far from population centers to serve as convenient landfills or are formed of matrix materials that are just too permeable. Most sanitary landfills have to be dug. If a hole is indeed excavated, the soil is saved to use for the daily cover. Whatever the origin of the cavity, today it will usually be lined before it goes into service, most often with several feet of dense clay and then with thick plastic liners made of strips that have been hot-sealed together. (Because regulations to this effect are recent, two thirds of all sanitary landfills—primarily the oldest ones—do not currently have liners.) When the liner is in place it is covered by several feet of gravel or sand.

Landfills produce a watery potage that drools to the bottom, and this leachate, as it is called, has to be anticipated and dealt with. The bottom contours of the newest generation of sanitary landfills are designed so that fluid, be it rainwater or Lemon Fresh or Budweiser or Olde English furniture polish, will flow toward drains through which perforated pipes have been threaded. The collected liquid is dealt with in a variety of ways, depending on whether the landfill operator subscribes to the "wet-landfill" or "dry-landfill" theory. The wet-landfill theory, which these days is adhered to by a tiny minority, holds that landfills should be saturated with as much liquid as possible in order to promote bacterial growth and biodegradation. Leachate is collected, sometimes treated, and pumped back to the top of the landfill, and thus is constantly recirculating; among other things, it is hoped that much of what is harmful in leachate will be absorbed or degraded as it percolates through fresh dry garbage. Landfills of this kind are illegal in most states and may well soon be extinct.

The dry-landfill theory—the one that the Environmental Protection Agency currently prefers—begins from the assumption that the drier a landfill is the less risk it poses of contaminating ground water. In dry landfills the leachate is collected and most of it does get treated. Some landfills have their own treatment plants; they treat the leachate like sewage, separating out the water, which is purified and released, and either dumping the solid sludge back into the landfill or burning the sludge and dumping the ash back into the landfill. Most landfills, however, send the leachate to the local municipal sewage facility. There, several things can happen to it. The water, of course, is always separated out, cleaned to regional standards, and released into a local river or the ocean. The rest is turned into sludge. The sludge is either dumped in the ocean, dumped in a landfill, burned, or used as fertilizer. If the content of the leachate is deemed hazardous, it is subject to the usual slew of regulations, and is eventually sent, at great expense, to a Subtitle C hazardous-waste disposal site.

Another substance that landfills can be counted on to produce is methane gas, a byproduct of decomposition. At many new landfills in the United States, within about five years of opening, augers will have drilled holes into the accumulated deposits from surface to bottom. Perforated pipes are inserted into the holes to draw off the methane, and are surrounded by gravel. At some landfills the methane gas sucked into these pipes is simply burned off or released into the atmosphere. At others the vents are connected to a storage station, where the methane is purified and then used to generate power locally or sold as fuel. With the help of engineering maps and accurate measurements of elevation an experienced bucket-auger handler, like the Garbage Project's Buddy Kellett, can drill a methane well with great precision, stopping the bit's advance inches before it pierces the lining, which would compromise the landfill's environmental integrity. (Piercing the liner is the landfill equivalent of a surgeon's accidentally perforating the gastrointestinal tract; neither mistake is necessarily irreparable, but neither should happen.) The operators of some of the newest landfills have begun setting a lattice of methane pipes into place before any garbage has even been dumped. The first pipes to be installed are short ones, and over time the pipes are extended, growing in height at the same pace as the landfill.

The daily tipping of garbage into a landfill is an orchestrated mechanical pavane that may begin as early as midnight (Fresh Kills runs twenty-four hours a day), but more usually starts at around 5:30 in the morning, when big mother-hen packer trucks or rigs pulling rectangular packer rolloffs from transfer stations file in noisily and deposit their cargoes across that day's "open face," in rows of piles, each tens of feet long and ten to twenty feet high. The piles are laid either on the top rim of the existing garbage glacier or in front of the bottom edge of the garbage pack—that is, either on top of or directly in front of the previous day's garbage. Next, bulldozers and machines called compactors that have five-foot-wide studded rollers push or squash the fresh, supple garbage into tight communion with the dirt-covered and somewhat more wilted deposits of the day before. By early afternoon all the garbage from a single day —a "cell" in the jargon of many landfill operators, although the terminology is not universal—has been pressed into place. From the side, the row upon row of cells looks like an arrangement of dominoes on their sides, leaning one against the other as if frozen at the moment of mid-collapse. As the garbage trucks become less frequent, special double-jointed vehicles with bays for bellies crawl up the dirt mounds near the garbage pit, fill up, rumble over to the latest cell, drop their loads, and return for another bellyful. Bulldozers coax the dirt so that it neatly covers the garbage.

In a typical landfill a cell is about twenty to thirty feet thick, twenty to twenty-five feet high, and a hundred feet long. Day by day, cell by cell, garbage spreads across the floor of a new landfill until it hits the far side. At that point a new layer—known as a "lift"—is begun. As a landfill's lifts accumulate, slopes and contours are shaped according to preplanned engineering specifications in order to direct rainfall runoff, give access to trucks and earth-moving equipment, and keep garbage avalanches at bay. Even after the final cap is bulldozed into place (the cap, which lies atop an especially thick stratum of dirt cover, is typically made of the same clays used to line the bottom of the landfill, thus helping to deflect rainwater around the whole structure) and the landfill is officially closed, the site will continue to produce methane gas for another fifteen to twenty years, and methane wells therefore must continue to operate. Nevertheless, soon after closure most contemporary landfills are landscaped and developed, and embark on second careers as golf courses, parks, or industrial estates, with only the methane wellheads, poking up like periscopes, to hint at the location's previous identity. In three or four decades nothing but the wellheads on the ground's surface will suggest to passersby the broken tricycles and crushed cereal boxes and millions of newspapers that lie underfoot. The amount of land that has been "recovered" during the past couple of centuries from landfills and other garbage repositories is extraordinary. The present contours of virtually every portion of New York City and the neighboring parts of New Jersey and Long Island have all been shaped by fill, much of it garbage (see Figure 4-A). Few people today have very much awareness that the local landfill is destined for an afterlife, or that many landscapes they take for granted conceal distinctly checkered pasts.

In many respects, then, our own civilization carries on the tradition passed along by previous ones: Rather than being buried by our garbage, we are rising above it. Modern sanitary landfills are expen-

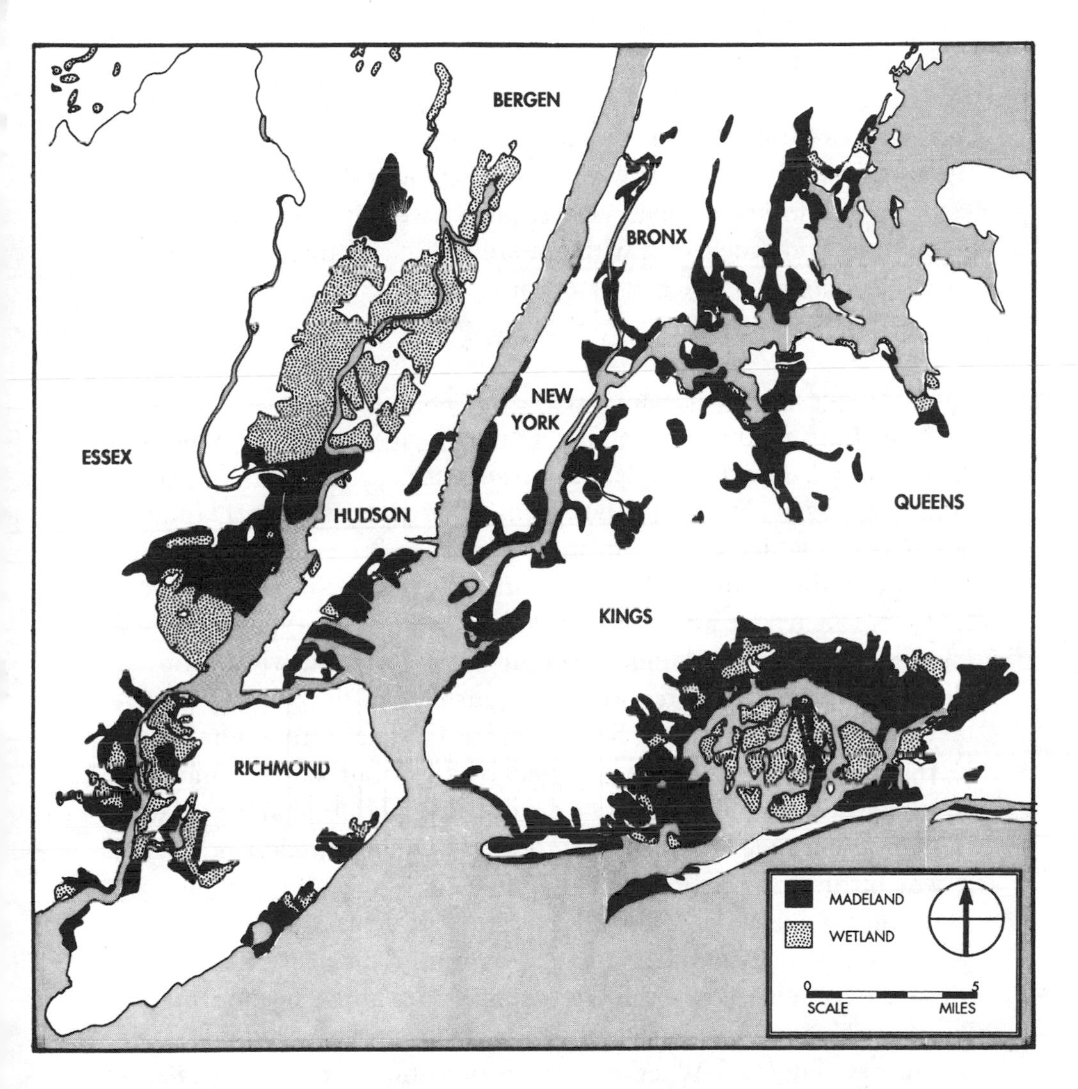

Figure **4-A.** The shaded sections represent those parts of the New York metropolitan area—former wetlands, in many cases—that, as of 1966, had been built up into solid land out of various kinds of debris, including large amounts of municipal solid waste.

SOURCE: *Waste Management*, Regional Plan Association, New York, 1968

sive to build—the construction of an eighty-acre landfill (which at present generation rates would serve a community of 500,000 for twenty years) would cost about $33 million, and the cost of closing the landfill when it was filled would be another $8 million. Two facts must be borne in mind. First, there can be no such thing as a world without landfills. They are an inevitable part of any conceivable garbage-disposal regime. Recycling and incineration, for example, both result in the production of wastes that must be landfilled. Second, new landfills are better sited and better designed than old ones. They may not be the most welcome of neighbors, but when we, as a society, decide not to open new landfills, we have also decided, by default, to continue living with the landfills that already exist, some of which may be problematic in character.

The Garbage Project began excavating landfills primarily for two reasons, both of them essentially archaeological in nature. One was to see if the data being gleaned from garbage fresh off the truck could be cross-validated by data from garbage in municipal landfills. The second, which derived from the Garbage Project's origins as an exercise in the study of formation processes, was to look into what happens to garbage after it has been interred. As it happens, the first landfill excavation got under way, in 1987, just as it was becoming clear—from persistent reports about garbage in the press that were at variance with some of the things the Garbage Project had been learning—that an adequate knowledge base about landfills and their contents did not exist. It was during this period that news of a mounting garbage crisis broke into the national consciousness. And it was during this period that two assertions were given wide currency and achieved a status as accepted fact from which they have yet to be dislodged. One is that accelerating rates of garbage generation are responsible for the rapid depletion and present shortage of landfills. The other is that, nationwide, there are few good places left to put new landfills. Whether these propositions are true or false—they happen, for the most part, to be exaggerations—it was certainly the case that however quickly landfills were being filled, the public, the press, and even most specialists had only the vaguest idea (at best) of what they were being filled up *with*. Yes, think tanks and consulting firms have done some calculations and come up with estimates of garbage quantities by commodity, based on national production figures and assumptions about rates of discard. But until 1987, when the Garbage Project's archaeologists began systematically sorting through the evidence from bucket-auger wells, no one had ever deliberately dug into landfills with a view to recording the inner reality in minute detail.

The Garbage Project was not without some slim archaeological precedent, which dates back to the summer of 1921. While writing up his now-famous dig at Pecos Ruin, on the headwaters of the Pecos River in San Miguel County, New Mexico—a study based on stratigraphic excavation techniques, which established the culture sequence among native peoples in the American Southwest—the pioneering archaeologist Alfred Vincent Kidder worked at Phillips Academy, in Andover, where he was a member of the department of archaeology. Kidder, the first American archaeologist to recognize the significance of stratigraphic layers in ancient ruins and ancient rubbish, became intrigued by a large trench that was being cut through the town of Andover's garbage dump to hold a multicommunity sewer pipe, and he spent a considerable amount of time at the work-site, down in the trench. He was able to see clearly in the strata the transition in light fixtures from whale-oil lamps to light bulbs. He was much taken with Milk of Magnesia bottles, because unlike many bottles the brand name was embossed on the glass, making for easy identification. Just about all archaeological excavations turn up objects whose purpose cannot be determined (these objects, it sometimes seems, always end up being thrown into the catchall category "religious paraphernalia"), and the Andover dig was no exception: Kidder found a large number of mysterious pieces of flat, rusted iron, some twelve to fourteen inches long. "I couldn't imagine what they were," Kidder would later write. "I took one of them and Madeleine [Kidder's wife] didn't know what they were, and I showed them to my mother, who was visiting us at the time. She said, 'Oh, those are corset bones. When your corset wore out we used to roll it up and tie it with a string and throw it in the rubbish.' They were made of metal. The whalebone ones had gotten to be so expensive that no one used them anymore."

Kidder's brief, serendipitous peek inside the Andover dump has become the stuff of archaeological lore—from the Garbage Project's point of view, it holds a status equivalent to Wilhelm Konrad Roentgen's serendipitous discovery of X rays, in 1895, at the Royal University of Wurzburg, or Alexander Fleming's accidental discovery of penicillin, in 1928, at St. Mary's Hospital, in London—but for more than six decades, strangely, no one followed Kidder's lead.

The first landfill excavated by the Garbage Project, in April of 1987, was the Vincent H. Mullins landfill, in Tucson (the landfill is named, appropriately, for a sanitation supervisor who in the early 1970s had delivered fresh garbage samples to Garbage Project crews). In the years since then, eight other landfills around the United States have been opened up and explored. The landfills were selected to represent varying climates and levels of rainfall, varying soils and geomorphology, and varying regional lifestyles; the garbage deposited in these landfills has been accumulating in some cases for more than forty years. As of mid-1991 the sample included two landfills in Arizona (Mullins in Tucson and the Rio Salado landfill in Tempe, both unlined; average annual rainfall, eleven inches; sandy soils used as cover; garbage deposited since 1952). There were two in California, at the southern end of San Francisco Bay (the Durham Road landfill, in Fremont, and the Sunnyvale landfill, in Sunnyvale, both unlined; average annual rainfall, twenty-three inches; gritty, loamy soils used as cover; garbage deposited since 1964). There were two in the Chicago suburbs (the Greene Valley landfill, in Naperville, and the Mallard North landfill, in Hanover Park, lined and unlined, respectively; average annual rainfall, twenty-nine inches; average annual snowfall, thirty-eight inches; dense clay soils used as cover; garbage deposited since 1970). There were two in the vicinity of Naples, Florida (the Collier County landfill, in the Everglades, and the Naples Airport landfill, on the south side of the airport, lined and unlined, respectively; average annual rainfall, eighty inches; sandy, loamy soils that must be trucked in used as cover; garbage deposited since 1974). And there was one in New York City (the Fresh Kills landfill, unlined; average annual rainfall, forty-three inches; average annual snowfall, twenty-eight inches; no soil cover used because the landfill is in operation twenty-four hours a day; garbage deposited since 1948). Additionally, in the pursuance of specific projects there have been limited excavations at two other

U.S. landfills, both in Tucson.* Several major excavations lie ahead. The fond ambition of the Garbage Project's staff is to be able one day to add to this list of excavated sites a garbage-dumping ground outside of London that has been in continuous use since at least the fifteenth century.

In terms of their environmental context, the differences among these landfills are extreme. In the Arizona desert the riverbeds are dry for three-quarters of the year, and then run in torrents during the late summer rainy season. In semitropical Florida, alligators sun themselves within sight of landfills and even bask in the leachate ponds. What is striking, however, is the extent to which the contents of these landfills seem to be relatively uniform from one part of the country to another. During its nine U.S. landfill excavations the Garbage Project retrieved 206 samples from sixty-five auger wells (up to eighty feet deep) and numerous backhoe trenches (dug to a depth of twenty-two feet), and exhumed a total of 28,426 pounds of garbage; the wells and trenches at each landfill were placed to ensure a representative sampling by date of refuse deposition. When commodity categories are compared from one landfill to another, the variance turns out to be negligible. For example, by weight the amount of rubber retrieved from the Mullins, Durham Road, and Greene Valley landfills fell in all cases at between 0.4 and 0.6 percent of the total weight of the refuse samples taken at each place. In all nine landfills textiles varied between 2.1 and 3.6 percent of refuse weight. The similarities extended to paper, plastic, and metals—indeed, to every category available. (Some of the slight differences that did exist, such as the somewhat lower proportion of paper in California's garbage than in that of Illinois, reflect different rates of recycling from place to place.) The lack of much variance is a reassuring indication that the Garbage Project's findings with respect to landfill content are dependable.

One key aim of the landfill excavations was to get some idea of the volume occupied by various kinds of garbage in landfills. Al-

* As noted in chapter one, four garbage sites in Canada have also been excavated, all of them in Ontario. They are the Burlington landfill, in Burlington; the Brock West landfill, in Pickering; the Oakville landfill, in Oakville; and the West Mall dump, in Etobicoke. A total of three tons of garbage was sorted at the four sites. Most of the data remain unevaluated.

though many Garbage Project studies have relied on garbage weight for comparative purposes, volume is the critical variable when it comes to landfill management: Landfills close not because they are too heavy but because they are too full. And yet reliable data on the volume taken up by plastics, paper, organic material, and other kinds of garbage once it has been deposited in a landfill did not exist in 1987. The Garbage Project set out to fill the gap, applying its usual sorting and weighing procedures to excavated garbage, and then adding a final step: a volume measurement. Measuring volume was not a completely straightforward process. Because most garbage tends to puff up with air once it has been extracted from deep inside a landfill, all of the garbage exhumed was subjected to compaction, so that the data on garbage volume would reflect the volume that garbage occupies when it is squashed and under pressure inside a landfill. The compactor used by the Garbage Project is a thirty-gallon cannister with a hydraulic piston that squeezes out air from plastic bags, newspapers, cereal boxes, mowed grass, hot dogs, and everything else at a relatively gentle pressure of 0.9 pounds per square inch. The data on garbage volume that emerged from the Garbage Project's landfill excavations were the first such data in existence.

What do the numbers reveal? Briefly, that the kinds of garbage that loom largest in the popular imagination as the chief villains in the filling up and closing down of landfills—fast-food packaging, expanded polystyrene foam (the material that coffee cups are made from), and disposable diapers, to name three on many people's most-unwanted list—do not deserve the blame they have received. They may be highly visible as litter, but they are not responsible for an inordinate contribution to landfill garbage. The same goes for plastics. But one kind of garbage whose reputation has thus far been largely unbesmirched—plain old paper—merits increased attention.

Over the years, Garbage Project representatives have asked a variety of people who have never seen the inside of a landfill to estimate what percentage of a landfill's contents is made up of fast-food packaging, expanded polystyrene foam, and disposable diapers. In September of 1989, for example, this very question was asked of a group attending the biennial meeting of the National Audubon Society, and the results were generally consistent with those obtained from surveys conducted at universities, at business meetings, and at conferences of state and local government officials: Estimates at the Audubon meeting of the volume of fast-food packaging fell mainly between 20 and 30 percent of a typical landfill's contents; of expanded polystyrene foam, between 25 and 40 percent; and of disposable diapers, between 25 and 45 percent. The overall estimate, then, of the proportion of a landfill's volume that is taken up by fast-food packaging, foam in general, and disposable diapers ranged from a suspiciously high 70 percent to an obviously impossible 125 percent.

Needless to say, fast-food packaging has few friends. It is designed to be bright, those bold reds and yellows being among the most attention-getting colors on a marketer's palette; this, coupled with the propensity of human beings to litter, means that fast-food packaging gets noticed. It is also greasy and smelly, and on some level it seems to symbolize, as do fast-food restaurants themselves, certain attributes of modern America to which modern Americans remain imperfectly reconciled. But is there really all that much fast-food packaging? Is it "straining" the capacity of America's landfills, as a 1988 editorial in *The New York Times* contended?

The physical reality inside a landfill is, in fact, quite different from the picture painted by many commentators. Of the more than fourteen tons of garbage from landfills that the Garbage Project has sorted, fewer than a hundred pounds was found to consist of fast-food packaging of any kind—that is, containers or wrappers for hamburgers, pizzas, chicken, fish, and convenience-store sandwiches, plus all the accessories, such as cups, lids, straws, sauce containers, and so on, plus all the boxes and bags used to deliver food and other raw materials to the fast-food restaurant. In other words, less than one-half of one percent of the weight of the materials excavated from nine municipal landfills over a period of five years (1985–89) consisted of fast-food packaging. As for the amount of space that fast-food packaging takes up in landfills—a more important indicator than weight—the Garbage Project estimate after sorting is that it accounts for no more than one-third of one percent of the total volume of a landfill's contents.

What about expanded polystyrene foam—the substance that most people are referring to when they say Styrofoam (which is a registered trademark of the Dow Chemical Corporation, and is baby blue in color and used chiefly to insulate buildings)? Expanded polystyrene foam is, of course, used for many things. Only about 10 percent of all foam plastics that were manufactured in the period 1980–89 were used for fast-food packaging. Most foam was (and is) blown into egg cartons, meat trays, coffee cups (the fast-food kind, yes, but mainly the plain kind that sit stacked upside down beside the office coffee pot), "peanuts" for packing, and the molded forms that protect electronic appliances in their shipping cases. All the expanded polystyrene foam that is thrown away in America every year, from the lowliest packing peanut to the most sophisticated molded carton, accounts for no more than 1 percent of the volume of garbage landfilled between 1980 and 1989.

Expanded polystyrene foam has been the focus of many vocal campaigns around the country to ban it outright. It is worth remembering that if foam were banned, the relatively small amount of space that it takes up in landfills would not be saved. Eggs, hamburgers, coffee, and stereos must still be put in *something*. The most likely replacement for foam is some form of coated cardboard, which can be difficult to recycle and takes up almost as much room as foam in a landfill. Indeed, in cases where cardboard replaced foam, it could often happen that a larger volume of cardboard would be needed to fulfill the same function fulfilled by a smaller volume of foam. No one burns fingers holding a foam cup filled with coffee, because the foam's insulating qualities are so effective. But people burn their fingers so frequently with plastic- or wax-coated cardboard coffee cups (and all cardboard hot-drink cups are coated) that they often put one such cup inside another for the added protection.

As for disposable diapers, the debate over their potential impact on the environment is sufficiently vociferous and complex to warrant its own chapter (see chapter seven). Suffice it to say for present purposes, though, that the pattern displayed by fast-food packaging and expanded polystyrene foam is apparent with respect to diapers, too. People *think* that disposable diapers are a big part of the garbage problem; they are not a very significant factor at all.

The three garbage categories that, as we saw, the Audubon respondents believed accounted for 70 to 125 percent of all garbage actually account, together, for only about 3 percent. The survey responses would probably have been even more skewed if respondents had also been asked to guess the proportion of a typical landfill's contents that is made up of plastic. Plastic is surrounded by a maelstrom of mythology; into the very word Americans seem to have distilled all of their guilt over the environmental degradation they have wrought and the culture of consumption they invented and inhabit. Plastic has become an object of scorn—who can forget the famous scene in *The Graduate* (or quote it properly)?—no doubt in large measure because its development corresponded chronologically with, and then powerfully reinforced, the emergence of the very consumerist ethic that is now despised. (What Mr. McGuire, a neighbor, says to Benjamin Braddock is: "I just want to say one word to you. Just one word. Are you listening? . . . Plastics. There is a great future in plastics. Think about it.") Plastic is the Great Satan of garbage. It is the apotheosis of the cheap, the inauthentic; even the attempts to replace or transform plastic—such as the recent ill-fated experiments with "biodegradable" plastic, which will be discussed in chapter seven—seem somehow inauthentic.

There are legitimate causes for concern about plastic, particularly with respect to its manufacture. For the moment the issue is the volume of plastics in landfills. Two statistics have received wide circulation. The first, which appears repeatedly in the press, is that while plastics may make up only 7 percent of all municipal solid waste by weight, they make up some 30 percent of municipal solid waste by volume. This 30 percent figure has a history: It comes from a report published by, and available (for $300) from, the International Plastics Consultants Corporation (IPCC), based in Stamford, Connecticut, a group that was set up to promote the recycling of plastic. The IPCC's methodology for estimating the volume in landfills occupied by plastics begins by accepting the Franklin Associates' materials-flows assumptions and their weight data on various garbage categories. To estimate the volume of various categories of garbage after such garbage has been crushed and compacted, the researchers obtained from the pertinent trade associations and businesses whatever data they had on the bulk density (that is, the volume per unit weight) of items that have been squashed and baled for transport, usually for shipment to recycling facilities.

There were, of course, a few problems. While the bulk density of some types of paper items, such as newsprint and corrugated cardboard, could be evaluated with a certain precision, because these items get recycled and records are kept, the IPCC had to assume that the bulk density of nonrecycled paper items for which they had no data, such as cereal boxes, paper towels, and tissues, was the same as that of recyclable paper. Similarly, the IPCC had to assume that the bulk density of all nonrecycled plastics, from toothbrushes to tables, was the same as the bulk density for the kinds of recyclable plastic for which it had data—primarily PET (polyethylene terephthalate) plastic soda bottles, the kind that most soft drinks now come in. And, of course, there being no trade associations for yard waste, food waste, and many other kinds of garbage, the International Plastics Consultants Corporation had to settle for reasonable estimates of the bulk density of all these garbage categories. The IPCC ended up by concluding that plastics made up 27 percent of a typical landfill's contents, a figure that in news reports was then rounded up to 30 percent.

The second estimate that one encounters with some regularity for the volume of plastics in landfills is 20 percent. The provenance of this figure is a 1988 Franklin Associates study of landfill constituents by weight and volume. This figure is inflated because Franklin Associates (as its researchers readily admit) excluded the huge category "construction and demolition debris"—which accounts for about 12 percent by volume of a typical landfill's contents—from their estimation of the total landfill pie, thereby reducing the size of the pie and magnifying the relative proportions of the other constituents. The problem with construction and demolition debris, insofar as Franklin is concerned, is the same one faced by the IPCC: no one keeps records on it. There is no trade association for construction and demolition debris in Washington, and, because local communities are not normally responsible for collecting and carting away such debris, as they are other kinds of garbage, very often not even haphazard documentation exists. And besides, the federal government does not technically consider construction and demolition debris to be municipal solid waste (though it ends up in municipal landfills). For these reasons construction and demolition debris was simply left out of the picture. By Franklin's account, not one ounce of construction and demolition debris—not one cinderblock, two-by-four, or rebar rod—has technically entered American landfills during the past thirty years.

The Garbage Project's methodology has not been quite as sophisticated as that of Franklin or the IPPC: Garbage Project personnel simply measured by weight and volume everything exhumed from sample municipal-solid-waste landfills. The results differ from the Franklin and IPCC numbers. In landfill after landfill the volume of all plastics—foam, film, and rigid; toys, utensils, and packages—from the 1980s amounted to between 20 and 24 percent of all garbage, as sorted; when compacted along with everything else, in order to replicate actual conditions inside a landfill, the volume of plastics was reduced to under 16 percent.

Even if its share of total garbage is, at the moment, relatively low, is it not the case that plastics take up a larger proportion of landfill space with every passing year? Unquestionably a larger number of physical objects are made of plastic today than were in 1970 or 1950. But a curious phenomenon becomes apparent when garbage deposits from our own time are compared with those from strata characteristic of, say, the 1970s. While the number of individual plastic objects to be found in a deposit of garbage of a constant size has increased considerably in the course of a decade and a half—more than doubling—the proportion of landfill space taken up by these plastics has not changed; at some landfills, the proportion of space taken up by plastics was actually a little less in the 1980s than it was in the 1970s.

The explanation appears to be a strategy that is known in the plastics industry as "light-weighting"—making objects in such a way that the objects retain all the necessary functional characteristics but require the use of less resin. The concept of light-weighting is not limited to the making of plastics; the makers of glass bottles have been light-weighting their wares for decades, with the result that bottles today are 25 percent lighter than they were in 1984. (That is why bottles in landfills are likely to show up broken in the upper, more-recent, strata, whereas lower strata, holding garbage from many years ago, contain many more whole bottles.) Environmentalists might hail light-weighting as an example of source reduction. Businessmen embrace it for a different reason: sheer profit. Using fewer raw materials for a product that is lighter and therefore cheaper to transport usually translates into a competitive edge, and companies that rely heavily on plastics have been light-weighting ever since plastics were introduced. PET soda bottles had a weight of 67 grams in 1974; the weight today is 48 grams, for a reduction of 30 percent. High-density polyethylene (HDPE) milk jugs in the mid-1960s had a weight of 120 grams; the weight today is about 65 grams, for a reduction of more than 45 percent. Plastic grocery bags had a thickness of 30 microns in 1976; the thickness today is at most 18 microns, for a reduction of 40 percent. Even the plastic in disposable diapers has been light-weighted, although the super-absorbent material that was added at the same time (1986) ensures that even if diapers enter the house lighter they will leave it heavier than ever. When plastic gets lighter, in most cases it also gets thinner and more crushable. The result, of course, is that many more plastic items can be squeezed into a given volume of landfill space today than could have been squeezed into it ten or twenty years ago.

This fact has frequently been met with skepticism. In 1989, Robert Krulwich, of the CBS network's "Saturday Night with Connie Chung" program, conducted a tour of the Garbage Project's operations in Tucson, and he expressed surprise when told about the light-weighting of plastics. He asked for a crushed PET soda bottle from 1989 and tried to blow it up. The light plastic container inflated easily. He was then given a crushed PET soda bottle found in a stratum dating back to 1981—a bottle whose plastic would be considerably thicker and stiffer. Try as he might, Krulwich could not make the flattened container inflate.

One item that has not been light-weighted during the past few decades is your typical daily newspaper—the messenger that repeatedly carries warnings about the garbage crisis. A year's worth of copies of *The New York Times*, for example, weighs about 520 pounds and occupies a volume of about 1.5 cubic yards. A year's worth of *The Times* is the equivalent, by weight, of 12,480 empty aluminum cans or 48,793 Big Mac clamshell containers. It is the equivalent, by volume, of 18,660 crushed aluminum cans or 14,969 crushed Big Mac clamshells.

Newspapers epitomize the part of the garbage problem that gets the least amount of attention: paper. During the 1970s futurists and other writers, perceiving the advent of an electronic society, heralded the new paperless workplace, the new paperless culture. "One of the most startling features of the Computer Revolution," Christopher Evans wrote in *The Micro Revolution* (1979) "is that print and paper technology will appear as primitive as the pre-Caxtonian handcopying of manuscripts seems to us. In sum, the 1980s will see the book as we know it, and as our ancestors created and cherished it, begin a slow but steady slide into oblivion." Predictions like that one were never quite believable even in their heyday, when the consequences of the advent of copying machines were already apparent. It is obvious by now that computers, far from making paper obsolete, have made it possible to generate lengthy hard-copy documents more easily than ever before. A computer with a printer is, in effect, a printing press, and there are now fifty-five million of these printing presses in American homes and offices, where twenty years ago there had been only typewriters. With respect to paper, advancing technology is not a contraceptive but a fertility drug. For one thing, as technology in general has become more and more sophisticated, with more and more components, the engineering specifications needed to describe complex systems have necessarily become more and more voluminous. One environmental consulting group recently publicized the assertion that if all the paper stored on a typical American aircraft carrier were removed, the ship would rise three inches in the water. Garbage Project researchers have been unable to substantiate that claim, but it is definitely the case that, prognostications to the contrary, paper has managed to hold its own among the components of the U.S. solid-waste stream. Edward Tenner, an executive editor at Princeton University Press, recently observed: "The paperless office, the leafless library, the inkless newspaper, the cashless, checkless society—all have gone the way of the Empire State Building's dirigible mooring, the backyard helipad, the nuclear-powered convertible, the vitamin-pill dinner, and the Paperwork Reduction Act of 1980."

For all the competition since the 1950s from plastic, metal, construction-and-demolition debris, and non-paperaceous organics, paper's contribution to a landfill's contents has remained relatively even, at well over 40 percent (see Figure 4-B). Newspapers alone

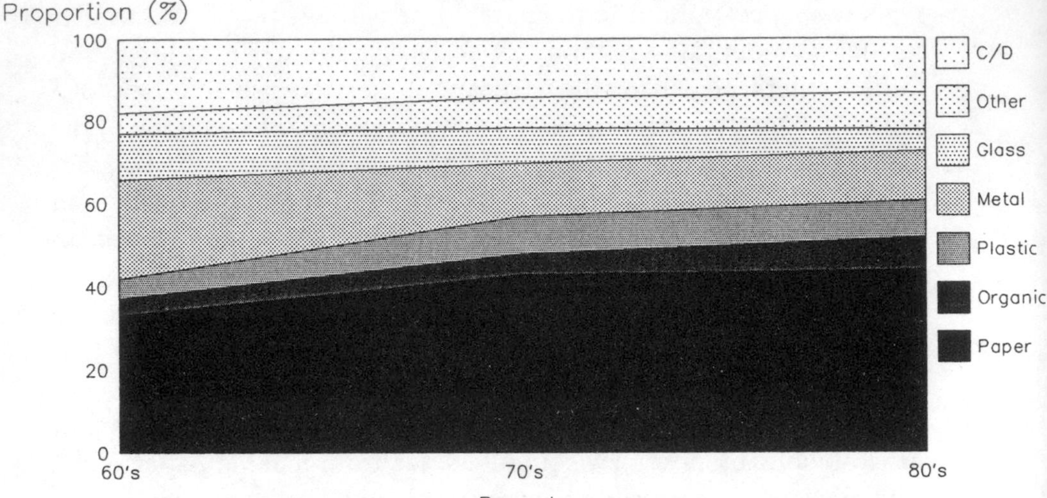

Figure **4-B.** Garbage Project excavations of landfills yield a picture of their changing composition over time; the data here reflect volumes that have been compaction-corrected. Paper is the single biggest constituent of a typical landfill. A voluminous but usually overlooked constituent of landfills is construction-and-demolition debris, which accounts for an average of about 12 percent of total content. The graph excludes soil used for cover.

SOURCE: The Garbage Project

may take up some 13 percent or more of the space in the average landfill—nearly as much as all plastics. Paper used in the packaging of consumer goods has grown in volume by about a third since 1960. Non-packaging paper—computer paper, stationery, paper plates and cups, junk mail—has doubled in volume. The volume of discarded magazines has likewise doubled, to about 1.2 percent—about as much as all the thrown-away fast-food packaging and expanded polystyrene foam combined.

One noteworthy contributor to a landfill's paper content is the telephone book. Dig a trench through a landfill and telephone books can be seen to stud some strata like currants in a cake. They are thrown out regularly, once a year; in the city of Phoenix, that means almost twelve pounds of phone books annually (one yellow pages and one white pages) for every business and household. And their expansion in number seems to know no bounds. First there are the normal "Baby Bell" phonebooks published by the seven regional phone companies, often two or three of them per household in a city of average size. Then come the many competing brands of Yellow Pages published by rivals to the Bell system companies: Reuben H. Donnelly and GTE Directories are the biggest, but there are some two hundred other yellow pages publishers. And then there are phonebooks that target specific businesses, or senior citizens, or juveniles, or members of different ethnic groups. Miniature, paperback book–sized phonebooks have recently appeared for people who have car phones, to ride beside them on the front seat. In most cases phonebooks are made of paper of such low quality that recycling is difficult, although some end uses do exist.

The avalanche of paper, like everything else about garbage, needs to be seen in perspective. Paper is not inherently a bad thing. There are many uses for paper that end up *limiting* the generation of garbage. The skillful packaging of food products, to give just one example, cuts down markedly on the wastage of foods. But for all paper's virtues, an inarguable fact remains: If garbage volume is ever to be significantly reduced, paper is the foe that must be faced. The task of getting some control over paper is made all the more necessary by the fact that paper and many other organics, as we will see in the next chapter, tend not so much to degrade in landfills as to mummify. They do not, in other words, take up appreciably less and less space as time goes by.

The following chart, which contrasts the findings of a 1990 Roper Poll with recent Garbage Project data, helps to summarize the difference between mental and material realities with respect to landfills. The percentages in the Roper column indicate the proportion of respondents identifying a particular item as a major cause of garbage problems.

	ROPER (%)	ACTUAL VOLUME IN LANDFILLS (%)
Disposable diapers	41	<2
Plastic bottles	29	<1

	Roper (%)	Actual Volume in Landfills (%)
Large appliances	24	<2
Newspapers	11	~13
All paper	6	>40
Food and yard waste	3	~7
Construction debris	0	~12

Misperceptions such as these are not harmlcss. They can lead to policies and actions that are counterproductive.

In commemoration of Earth Day, 1990, the New York Public Interest Research Group launched a campaign against the use of certain highly visible and famously odious forms of garbage, such as fast-food containers, aseptic packaging (juice boxes), and disposable diapers, and it urged members of allied environmental groups to spread the word "through newsletters and other publications." One can appreciate the good intentions—as well as the irony of the means of communication employed.

Popular misconceptions about what landfills are filled with are matched by popular misconceptions about how fast they are filling up. There can be no disputing the fact that there is, for the time being, an acute shortage of landfills still available to take deposits, especially in the northeastern United States. Since 1978, according to the Environmental Protection Agency, some fourteen thousand landfills have been shut down nationwide (leaving some six thousand in operation). Still, as the University of Pennsylvania's Iraj Zandi has shown, these figures do somewhat overstate the problem—and even the EPA is half-hearted about offering them. Many of the shut-down "landfills" were actually open dumps being closed for environmental reasons, and whatever the nature of the sites, they have tended to be relatively small, whereas those that remain open are quite large. In 1988, for example, 70 percent of the nation's landfills—the smaller ones—handled less than 5 percent of the municipal solid waste that was landfilled; that same year, fewer than five hundred landfills, or about 8 percent of the total—the bigger ones—handled nearly 75 percent of the nation's landfilled garbage. "It appears," Zandi writes, "that the trend is toward operating fewer but larger landfills. This phenomenon coincides with the trend in the rest of the industrialized world." (As of 1990, some 42 percent of all landfills were under ten acres in size, 51 percent were between ten and 100 acres in size, and 6 percent were larger than 100 acres. Most new landfills being created are of the large variety.)

That said, the situation regionally is in many cases dire. In New Jersey, the number of landfills has dropped from more than three hundred to about a dozen during the past fifteen years, and more than half of New Jersey's municipal solid waste must now be exported to landfills in other states—for the most part, states in the Midwest, whose many depressed rural counties and private landfill owners are willing to take the money that comes with the garbage, even if the relentless convoys of eighteen-wheel tractor trailers unnerve and anger local residents (see Figure 4-C). The customary formulation of the problem that we face (it appears in virtually every article on the subject) is that 50 percent of the landfills now in use will close down within five years. As it happens, that has long been the general state of affairs—it was true in 1970 and 1960—because the waste-management industry has never seen the need to maintain excess capacity beyond roughly that level. In the past, however, new landfill capacity was rarely hard to obtain. The difference today is that in many places used-up capacity is simply not being replaced. In 1976, for example, the state of Texas awarded some five hundred permits for landfills; last year the state awarded only fifty. The inevitable result in such cases is scarcity, ruinously high tipping fees (the amount that landfills charge customers for dumping garbage—upwards of $125 a ton in the most congested areas), and a desperate search by communities for alternatives.

Why are more permits not being granted? The reasons usually have nothing to do with the claim that one frequently hears: that we are running out of room for them. Yes, it is sometimes the case that we *have* run out of room. In the congested northeast there is not all that much space left for landfills, at least not safe ones. Some 1,350 twenty-ton tractor trailers laden with garbage now leave Long Island every day, bound for distant repositories. In the nation as a whole,

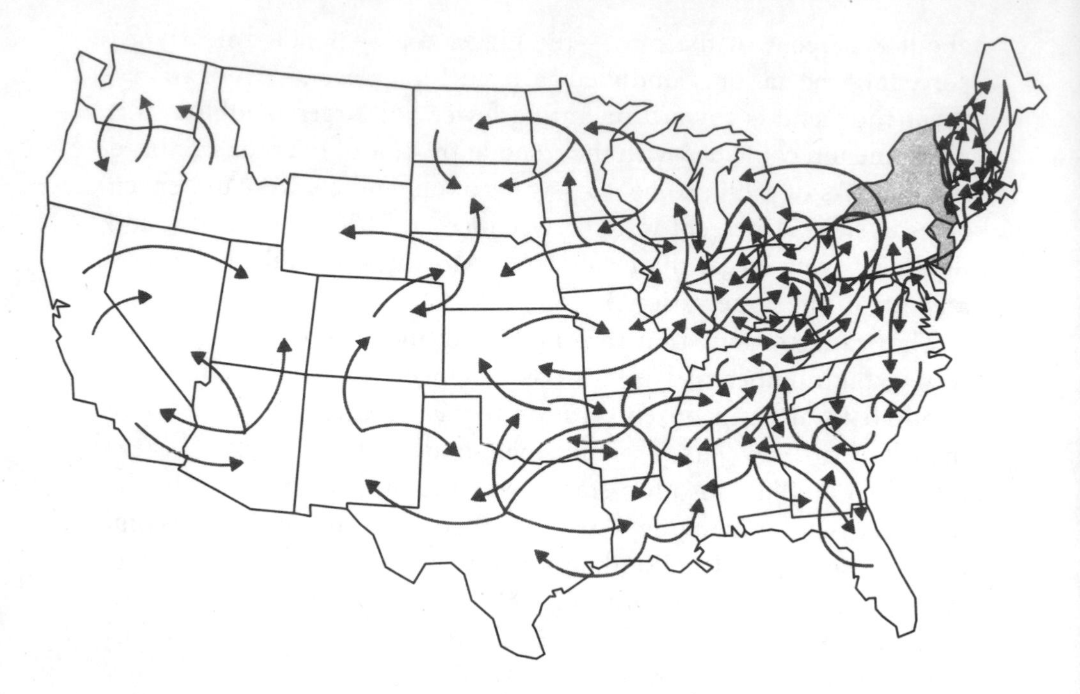

Figure 4-C. Interstate traffic in garbage has grown increasingly heavy and complex. In order to remain comprehensible, this map does not show the movement of garbage out of New York and New Jersey, the two biggest garbage-exporting states in the country. New York's garbage is trucked as far away as New Mexico.
SOURCE: National Solid Wastes Management Association

however, there is room aplenty. The United States is a big country, heavily urbanized but with enormous tracts of empty countryside. A. Clark Wiseman, in a study published by the Washington-based think tank Resources for the Future, has calculated that if the current rate of generation were maintained, all of America's garbage for the next one thousand years would fit into a landfill space 120 feet deep and forty-four miles square—a patch of land representing less than 0.1 percent of the surface area of the United States, or equivalent in size to three Oklahoma Citys. Such a landfill is for any number of reasons completely impractical, of course; the point here is simply that the total amount of space is not all that large. Few nations are as substantially endowed with uncongested territory as this one is, and there is appropriate land available even in relatively populous areas. Recently Browning Ferris Industries, one of the nation's two biggest full-service garbage disposal companies (the other is Waste Management, Inc.), commissioned an environmental survey of eastern New York State with the express aim of determining where landfills might safely be located. The survey pinpointed sites that constituted only 1 percent of the region's land area, but that still represented two hundred square miles of territory. And yet with all this potentially available land, the state of New York has since 1982 closed down 298 landfills and opened only six.

The obstacles to new sanitary landfills these days are to some extent monetary—as noted earlier, landfills are expensive—and, more important, psychological and political. Nobody wants a garbage dump in his or her neighborhood. The focus of NIMBY ("not in my back yard") protests is ostensibly community safety. In truth, however, problematic but existing landfills on inappropriate sites tend to draw less heat than well-planned but as yet only proposed landfills intended for appropriate sites. The key variables are property values and political clout. In metropolitan areas in particular, many existing landfills are to be found in socioeconomically depressed locations. Many new landfills, in contrast, are proposed not for congested metropolitan areas but for the far hinterland just beyond the older suburbs: in the heartland of the exurban gentry. The inhabitants are people who have the money, the knowledge, and the will to fight. And few politicians see making a principled case for a local landfill as the way to further their careers.

These are real problems—landfills filling up, difficulties opening replacements—and as so often seems to be the case with this nation's intractable ills, the only meager solace one can find is in the fact that they are nothing new. "Appropriate places for garbage are becoming scarcer year by year. . . . Already the inhabitants in proximity to the public dumps are beginning to complain." Those words were written by the chief health officer of Washington, D.C.—in 1889.

CHAPTER 5

THE MYTHS OF BIODEGRADATION

On a still October evening in 1986 the pop singer Steve Winwood gave a concert for a crowd of 11,200 at the Shoreline Amphitheater in the city of Mountain View, California, which lies forty miles south of San Francisco. At about 8:15 a member of the audience pulled a cigarette and plastic disposable lighter from his pocket and began to light up. A flick of the thumb produced a spark—and then a pillar of flame that shot five feet into the air, singeing the hair of a woman, Cheryl Ann Bogue, who was sitting nearby. Bogue was not seriously injured, but she filed a claim against the city of Mountain View, charging negligence. On what grounds did she base her claim? On these: Mountain View had built Shoreline Amphitheater on top of the city's old landfill; methane gas escaping from the landfill had been responsible for turning the lighter's winsome flicker into a flamethrower.

We have all heard stories like this before: They are part of the folklore—mostly the urban folklore—of late-twentieth-century America. In truth, it isn't very often that a methane buildup in an old dump or landfill causes any sort of real problem (though the incidents that do occur tend to be widely publicized and, inevitably, have considerable influence on public perceptions). The Garbage Project once surveyed a variety of landfill-gas companies and other organizations involved in landfill operations with a view to amassing a file of as many examples of methane fires and explosions at landfills as possible. The effort didn't amount to much, because there have been very few such fires and explosions. Why? Part of the reason is that newer landfills have been built with methane-venting or methane-collection systems in place, and some of the older ones have been retrofitted with such systems. (In the wake of the incident at the Steve Winwood concert, the city of Mountain View decided to spend $2.5 million to retrofit the Shoreline site with a methane-collection system.) But methane wells exist as yet at only a minority of landfills, suggesting that there must be other reasons for the infrequency of dangerous leaks—reasons that perhaps involve the internal dynamics of garbage deposits themselves.

Misconceptions about the interior life of landfills are profound—not surprisingly, since so very few people have actually ventured inside one. There is a popular notion that in its depths the typical municipal landfill is a locus of roiling fermentation, of intense chemical and biological activity. That perception is accompanied by a certain ambivalence. A landfill is seen, on the one hand, as an environment where organic matter is rapidly breaking down—biodegrading—into a sort of rich, moist, brown humus, returning at last to the bosom of Mother Nature. Biodegradation, in this view, is something devoutly to be desired, an environmentally correct outcome of the first order, perhaps even part of God's plan. Romantic thinking about biodegradation is widespread. It lies behind such dubious ventures as the proposed development by the British company London International of a biodegradable latex condom. On the other hand, coexisting with the romance of biodegradation, there is the view of a landfill as an environment from which a toxic broth of chemicals leaches into the surrounding soil, perhaps to pollute groundwater and nearby rivers and lakes. What both views of landfills have in common is the assumption that a great deal of biodegradation is taking place.

Some biodegradation *is* taking place—otherwise landfills would produce none of the large amounts of methane, or of the trace emissions of benzene, hydrogen sulfide, chlorinated hydrocarbons, and other gasses, that they do in fact produce. The truth is, however, that the dynamics of a modern landfill are very nearly the opposite of what most people think. Biologically and chemically, a landfill is a much more static structure than is commonly supposed. For some kinds of organics, biodegradation goes on for a little while, and then slows to a virtual standstill. For other kinds, biodegradation never really gets under way at all. Well-designed and managed landfills seem to be far more apt to preserve their contents for posterity than to transform them into humus or mulch. They are not vast composters; rather, they are vast mummifiers. Furthermore, this may be a good thing. For while there are positive things to say about biodegradation, the more of it that occurs in a landfill, the more opportunities there will be for the landfill's contents to come back to haunt us.

When the Garbage Project set up shop, in 1972, its focus was not on the garbage crisis, and, as noted, it did not begin by excavating landfills. The initial emphasis was on gaining insights into people's behavior, and the garbage examined was fresh off the truck. Although the invasive odor of fresh garbage at least hinted that some degree of putrefaction was probable, the issue of biodegradability was not addressed. Project members simply assumed—like everyone else—that widespread biodegradation was the inevitable lot of the organic material dumped into landfills.

In hindsight, it is clear that clues to the actual state of affairs existed long before the Garbage Project undertook extensive investigations. One clue involved a report by an environmental consulting firm which noted that although more than half of all municipal solid waste consists of materials that are at least in theory biodegradable, for some reason, even twenty or thirty years after being closed, most landfills have settled no more than a few feet at most. Another clue involved data on landfill methane production, which in most cases amounts to no more than 50 percent of what it theoretically should be, and in some cases amounts to as little as one percent. A third clue was an account by the archaeologist Rodolfo Lanciani, published in 1890, of his excavation at an ancient Roman garbage dump on the Esquiline Hill, which revealed that much of the garbage from imperial times had yet to fully decompose. "On the day of the discovery of the above-mentioned stone, June 25, 1884," Lanciani writes, "I was obliged to relieve my gang of workmen from time to time, because the smell from that polluted ground (turned up after a putrefaction of twenty centuries) was absolutely unbearable even for men so hardened to every kind of hardship as my excavators."

The true state of affairs revealed itself when the Garbage Project's research priorities began shifting increasingly to public-policy issues involving garbage—and the research venue began shifting increasingly to landfills. Instead of garbage that was at most a few days old, researchers began dealing with garbage that was ten, twenty, thirty years old—sometimes even older. Various artifacts began to accumulate in the Project's storage bins, particularly in the form of old newspapers with intriguing or resonant headlines: "Apollo Orbits Moon," July 30, 1971 *(Arizona Republic);* "Customs Men Bar Hippies to Cut Mexican Dope Flow," October 18, 1967 *(Phoenix Gazette);* "40 Red MIGs Downed or Hit During Week," April 5, 1952 *(Phoenix Gazette);* "Hint Dropped by Truman Suggests He May Not Be Candidate for President," January 8, 1952 *(Tempe Daily News).* As noted earlier, newspapers are extremely valuable for the purpose of dating garbage deposits, and they were of course entertaining curiosities in their own right. In the tradition of "The Emperor's New Clothes," though, it took a visitor to one landfill excavation—at the Mallard North landfill, in Elgin, Illinois—to point out the obvious. Casting his eyes one day in June of 1988 over the ranks of sorting bins holding stacks and stacks of old newspapers, he said: "I thought newspapers were supposed to biodegrade." As if to reinforce the point, Mallard North, as it happens, is the landfill that yielded up that fifteen-year-old steak, its bone, fat, and lean in a lot better condition than Ramses II (and without benefit of embalming).

Once broached, the subject of biodegradability became the target of a major research program. The first question to answer was: After a period of ten or fifteen years, how much paper and other organic garbage remains in landfills; that is to say, how much does not become transformed into methane and humus? There is, of course, some variability from landfill to landfill, but when the volume of paper items is combined with those of food waste, yard waste, and wood (mostly lumber used in construction), the overall volume of organic material recovered from the nine U.S. landfills excavated by the Garbage Project is extraordinarily high. For example, organics represented 32.5 percent of the ten- to fifteen-year-old garbage excavated at the Naples Airport landfill, 50.6 percent of the garbage of the same age excavated at Mallard North, and 66.5 percent of the garbage of that age at Rio Salado, in Phoenix. Organics in four twenty- to twenty-five-year-old samples from the landfill at Sunnyvale, California, represented some 40 percent of the sampled garbage. Organics in four Rio Salado samples from the 1950s accounted for 49 percent of the samples' total volume. Almost all the organic material remained readily identifiable: Pages from coloring books were still clearly that, onion parings were onion parings, carrot tops were carrot tops. Grass clippings that might have been thrown away the day before yesterday spilled from bulky black lawn and leaf bags, still tied with twisted wire but ripped open by garbage trucks and landfill bulldozers. Whole hot dogs have been found in the course of every excavation the Garbage Project has done, some of them in strata suggesting an age upwards of several decades.

The percentages just cited need to be put into perspective: To get some sense of the pace of biodegradation in a landfill one needs to know not only the volume of organics still there after fifteen or twenty years but also the volume that was originally present. The Garbage Project's Douglas Wilson came up with a way of making the comparison. When the Mullins landfill, in Tucson, opened for business in 1979, the garbage it accepted came from certain specific parts of town. As it happens, from 1974 onward the Garbage Project had been sampling, sorting, and weighing fresh garbage from one of those parts of town. Comparing the proportional-composition data already compiled on the fresh garbage over a period of, say, seven years (1979–1986) with data on excavated garbage from strata at Mullins representing those same years could be expected to yield the sought-for information.

And it did. The first item that was looked for in the fresh-garbage/landfill-garbage comparison was plastic. Because plastic virtually does not biodegrade or otherwise change after burial, except to break apart, the same amount of it ought to be present after ten years as on the day it was dumped. The weight data on plastics—as a percent of total sample weight—indicated a close match between their proportion in the samples of fresh garbage and in the samples of excavated garbage (excluding construction and demolition debris). The fact that this was the case was an early tipoff to what was coming, for statistically the percentage of plastic in the landfill samples would have been somewhat higher than that in the fresh-garbage samples if significant amounts of other types of landfill garbage had disappeared through biodegradation.

The picture became even clearer when the percentage of paper by weight in the two data sets was examined, and the aging landfill sample turned out to be still well within the range of the fresh-garbage sample. Paper, in other words, was not biodegrading rapidly at all. Neither was lumber, as you might expect. As for other kinds of organic waste, such as food and yard waste, the comparison revealed that after five years or so its percentage by weight in landfill garbage had declined relative to its percentage by weight in fresh garbage; this drop-off continued for some years, and the gap in this category between landfill and fresh garbage therefore widened.

Organics like food and yard waste, then, were the only items that could be considered truly vulnerable to biodegradation under normal landfill conditions, and they accounted only for between 10 and 20 percent of the organic material in landfills and for only between 5 and 10 percent of total landfill contents. The evidence from excavations indicates that even after two decades of burial about one-third to one-half of these vulnerable organics remain in a recognizable condition. This portion continues to experience biodegradation thereafter, but probably at a snail's pace. Given these findings, it was gratifying to come across the results of an earlier experimental attempt by the Department of Civil and Environmental Engineering at the University of Cincinnati to speed up the biodegradation process by grinding up organic garbage before depositing it in a simulated landfill. The scientists reported: "Only a small fraction of the total ground refuse mass was decomposed after five years in the test cells. This fraction was believed to be the food wastes."

As we will see, the chemistry of biodegradation inside a landfill is a highly complex and problematic process; among other things, to the extent that biodegradation does occur, the activity is highly variable from place to place throughout the landfill. The puzzle hasn't been completely put together yet, though already pieces have been discovered that just don't seem to fit. Perhaps evidence will one day turn up that will prompt a reassessment of the information about biodegradation in these last few paragraphs. Certainly the pattern that has been revealed through archaeological excavations of landfills so far accords with what is known of the typical life cycle of a field of methane wells: They vent methane in fairly substantial amounts for fifteen or twenty years after the landfill has stopped accepting garbage, and then methane production drops off rapidly, indicating that the landfill has stabilized. Henceforward, it would seem, the landfill won't be changing very much at all.

Why doesn't a lot of garbage in landfills biodegrade? Biodegradation is a process that occurs all around us—effortlessly, it would seem—and the kinds of organic materials that end up in landfills are certainly capable of undergoing this process. The microorganisms responsible for the various kinds of biodegradation are not shy, nor are they few in number: It is estimated that the biomass of the microorganisms that live in the first several feet below the surface in a typical acre of ground comes to about half a ton. We see the handiwork of these microorganisms in the meanest compost pile behind the barn or the garage. We see their handiwork every day in parks and fields, in city streets and country lanes: the forlorn, discarded remnants of food and packaging and newspaper well along their way on the journey from utility to nullity.

In the laboratory, microbiologists have no trouble at all coming up with ways to promote biodegradation. Organic garbage is placed in a blender and ground into pieces that are about two square millimeters in size. Water is added to this mixture in quantities sufficient to boost the aggregate moisture content, often by as much as 200 percent. The container holding all this is sealed, with monitors puncturing the cap to measure the production of methane and other gasses. The temperature is maintained at an ideal level. Every day a laboratory assistant gives the container a good shake. Under these conditions the physical metamorphosis of garbage into gas (on the one hand) and a kind of soupy gray slime whose constituents can no longer be recognized (on the other) is relatively rapid: The process takes no more than a few months.

The problem is that laboratory conditions and even conditions in a compost pile—or in a field or city street—are usually not comparable to the conditions in a landfill. Biodegradation works most efficiently under composting conditions, when debris is chopped up, regularly turned, kept wet, and exposed to the oxygen that aerobic microorganisms, which biodegrade organic material in the most straightforward way, require. These conditions are not met in modern landfills. The garbage stays where it has been dumped, tightly compacted but largely intact. Although some 200 landfills do recycle leachate through their garbage deposits, as noted earlier adding fluid waste or other kinds of fluid to landfills is widely discouraged, for fear of increasing the possibility that toxic liquids will migrate. And below all but the very top layers of a landfill (about eight feet) microorganisms that require oxygen seem to survive in insignificant numbers.

Anaerobic microorganisms do exist in this oxygenless environment, of course, but the manner in which they achieve biodegradation is relatively complex. In the case of paper, for example, first clostridia or other bacteria must produce enzymes called cellulases, which break cellulose into smaller molecules, such as various sugars. Acetogenic bacteria must then ferment the sugars, producing the organic acid acetic acid. In the final step of biodegradation, bacteria known as methanogens must convert the acetic acid into methane. How quickly all this happens—or whether it happens at all—depends on many variables, such as the acidity and temperature of the landfill environment, and the particle size of the garbage itself.

Of all the landfill excavations that they have conducted, Garbage Project researchers have on only two occasions encountered gray slime resembling the slime in laboratory jars—and on only one of these occasions did the slime actually turn out to be biodegraded garbage. The first encounter occurred in June of 1988 at Mallard North, where a bucket auger taking a sample at a depth of about twenty feet returned to the surface carrying a kind of treacly gray lava studded with a few bottles and splinters and chunks of other objects. The response to Wilson Hughes's initial reaction—"We're not going to analyze this stuff, are we?"—came in the affirmative. The sample conceivably represented the first discovered example of extensively biodegraded garbage inside a municipal-solid-waste landfill. The Garbage Project crew dutifully assembled an array of jars and plastic bags, and preserved every foul, precious drop they could. The thoughts of very few at the time were centered on the inherent environmental friendliness of biodegradation.

This momentous episode came to be seen in a different light sometime later, when laboratory analysis revealed that the gray slime was just plain city sewage sludge that had been deposited in the landfill. Dumping sludge (or other refuse that is not municipal solid waste) in a landfill is known as "co-disposal," and the practice is widespread. (Some 16 percent of all landfills are co-disposal sites.) The reason sludge is dumped in landfills, apart from the fact that sludge needs to be dumped somewhere, is to "jump start" the biodegradation process. The attempt had been in vain at Mallard North. When the pieces of landfill garbage that had been awash in the gray slime were separated out and cleaned, their identity as unbiodegraded organic matter became apparent: leaves, twigs, and grass clippings, primarily, along with a few pages of newspaper. The date on one of the pages was March 23, 1972. The landfill has been closed since 1974.

It may be worth noting, in the context of Mallard North, that according to recent news reports, attempts to jump-start biodegradation by means of sewage sludge also proved to be in vain at the Richland County landfill, in South Carolina. This fact came indirectly to light when representatives of *The Greenville News* went looking through the landfill in 1990 for documents from the early 1980s that pertained to allegations of financial improprieties by the president of the University of South Carolina, James B. Holderman. The searchers, who were given a rough idea where to look by the man who had brought the documents to the landfill, and who were further assisted by having a rough date of deposit to serve as a homing device, dug exploratory trenches with a backhoe and finally found what they were looking for some ten feet below the surface. Though coated with sludge, which had been added to the landfill for the same reason it was added at Mallard North, the documents remained intact and legible. They filled twenty-five boxes, which were immediately impounded by state law-enforcement officials for possible use against Holderman, who was under indictment.

It was the search for a landfill where garbage really undergoes biodegradation that eventually led the Garbage Project to the Fresh Kills landfill, in New York City—and it was there, in the second encounter with gray slime, that the Project found what it was looking for. As recounted earlier, Fresh Kills landfill occupies a vast tract of what had been a tidal swamp. The wetlands were watered by a confluence of channels and streams, whose directional propensities continue to influence the flow of liquid under, through, and around the mounds of garbage that now command the site. Fresh Kills is also watered by the ebb and flow of the tides. At the time Fresh Kills received its landfill designation, in 1948, swamps were deemed to be the ideal landfill location—indeed, as late as 1973 one finds Katie Kelly reporting the view that landfills may be "a good way to fill in . . . marshes, sandy areas, and shorelines"—and plans called for a suburban community eventually to be built on all the "waste" land that garbage would help to reclaim on Staten Island. By the late 1950s, the entire Fresh Kills swamp, except for some of the larger channels, had been filled to the point that the surface of the landfill was level with the solid ground that had surrounded the swamp; the landfill was between twenty and forty feet deep. As other landfills in New York closed down, along with old incinerators, Fresh Kills shifted to a twenty-four-hour-a-day footing in order to keep pace with the fleet of garbage barges arriving daily from Manhattan and elsewhere in New York City. By the mid-1960s the landfill had in places risen considerably above the level of the solid ground that had surrounded the swamp, and the original plan to build a suburban subdivision in the neighborhood was shelved.

The bucket of gray slime that showered the Garbage Project crew at Fresh Kills was not sewage sludge, but the end product of biodegradation in this particular landfill environment. In the samples from the very deepest levels of virtually every well that was dug by Buddy Kellett's crew at Fresh Kills the gray slime contained no food debris or yard wastes, and only a few shreds of paper. (One bore the date July 7, 1949.) And the gray slime was not hot—it was 60 to 80 degrees Fahrenheit—meaning that the biodegradation process had ended. Samples from higher up in the landfill—from strata that can be correlated with the 1960s—showed a far more advanced state of biodegradation than samples of similar age from other landfills the Garbage Project has excavated. Some recognizable raw organic material was still visible—food and yard waste—but not very much, and the volume of paper was drastically diminished. At other landfills paper accounts for between 35 and 45 percent of total garbage volume in strata from the 1960s; in the comparable samples from Fresh Kills, it accounts for between 10 and 18 percent. Temperatures in these still-decomposing layers ranged from 100 to 140 degrees Fahrenheit, reflecting the continued biological activity. The Garbage Project crew eventually became adept at anticipating the arrival of a bucket of slime by simply monitoring the temperature trajectory of successive loads of garbage brought up by the bucket auger.

What makes Fresh Kills so different from other landfills excavated by the Garbage Project? It predates most other landfills in use today and, unlike many newer landfills, it is not lined: The garbage and planet Earth are in direct contact, unimpeded by any clay or plastic membrane. The point of contact, moreover, is a swamp: The environment is an exceedingly wet one, and wetness helps foment biodegredation. The high levels of moisture, it should be added, are not confined to areas where the landfill bottom is in direct contact with the wetlands. Garbage drinks up water like a sponge, with the result that the landfill is soaking wet several stories above the water level. Not only is the Fresh Kills garbage wet, a lot of the water is in a state of motion (simulating some of the movement of fluids in a compost pile): The tides sweep in and out of the garbage-laden wetland twice a day. This difference may be the critical one.

So: biodegradation occurs at Fresh Kills at rates not to be found —and for reasons not to be found—at the other landfill sites examined to date. Even at Fresh Kills, however, the biodegradation process has been numbingly slow—far slower than for an orange peel or a paper cup or a magazine simply left out by the side of the road. Scientists cannot yet fully explain why. As part of its preparations for the Fresh Kills excavation the Garbage Project was joined by a team of microbiologists and other specialists whose task it would be to scrutinize the landfill samples with a view to (among other things) shedding some light on the question of biodegradation. One environmental microbiologist, Anna Palmisano, of Procter & Gamble's Environmental Laboratories, in Cincinnati, analyzed twenty-eight samples of garbage from Fresh Kills (along with twelve from the Naples Airport landfill), looking specifically for evidence of anaerobic cellulose-degrading bacteria—the bacteria that take the first bite, so to speak, out of paper. Palmisano found many kinds of bacteria in the samples, all in rather high concentrations: millions per gram of garbage. But the distribution of cellulase enzymes was very patchy, and no cellulose-degrading bacteria were isolated at all.

Another microbiologist, Joseph Suflita, of the University of Oklahoma, has focused on the final step of the biodegradation of solid waste, in which methanogens and sulfate-reducing bacteria turn organic acids into gas. Suflita first became interested in sulfate degraders after observing the Garbage Project's Rio Salado excavation, in Tempe, Arizona. Suflita noticed that as most of the buckets came out of the ground much of the garbage was dark; over the next few hours—in some cases, over the next few days—the garbage gradually reacquired much of its original range of color. The phenomenon was indicative of the presence of significant amounts of sulfur or sulfur-related compounds. Samples of exhumed garbage from all garbage categories were tested for sulfur, and all yielded high sulfur values; however, while these samples were clearly serving as a reservoir of sulfur, they did not appear to be the actual source of the sulfur. Where had the sulfur come from? Suflita speculated that one source may have been leachate from the gypsum in wallboard, which is a major element of that "construction and demolition debris" category that so often gets overlooked. The significance is this: The presence of sulfur enhances the growth of sulfate reducers, and laboratory studies show that when sulfate reducers are active, they may compete successfully with methanogens for acetate. This, in turn, diminishes the capacity of the methanogens to do their job. Precisely how much biodegradation the sulfate reducers themselves can accomplish remains unknown.

Science will no doubt one day puzzle out the biochemistry of decomposition in landfills. For the time being the one thing we can be sure of, and thankful for, is that not an enormous amount of decomposition occurs in many landfills. Garbage, even household garbage, contains many substances that are best left alone. A landfill that is dry, quiet, and relatively inert, and that in a sense keeps to itself, is the kind of landfill that does civil engineers proud. A landfill that is wet, teeming with roisterous activity, and spilling its insides into the outside world, is the situation one wants to avoid. That way lies Fresh Kills, which pours at least a million gallons of leachate into New York Harbor every day.

The leachate issue can be broken down into two fundamental questions. First, is there really a lot of toxic material in a typical municipal landfill? Second, does that material travel? Since the beginning of time human beings have created artifacts that have posed certain risks to their makers, and the inhabitants of the present age are obviously no exception. Leaving aside the toxic chemicals and other waste that is generated and disposed of by industry, America's small businesses and almost every one of its households consume and discard countless items that contribute a steady flow of poisonous, carcinogenic, or otherwise hazardous substances into the municipal waste stream. Take nail polish: A typical half-ounce bottle contains xylene, dibutyl phthalate, tolulene, and other potential pollutants whose names appear on the Environmental Protection Agency's aptly named P- and U-lists. (The P- and U-lists cite chemicals often used in commercial production that are either highly ignitable, corrosive, reactive, or toxic; chemicals on the lists are known as "listed" hazardous waste.) If you regularly bought the chemicals in your nail polish in fifty-five-gallon drums instead of in a half-ounce bottle, you would be legally prohibited from discarding the chemicals in a municipal solid-waste landfill. You would be required, rather, to transport them to a state-licensed Subtitle C hazardous waste disposal site (the nearest ones to the Garbage Project's headquarters in Arizona are in Beatty, Nevada, and Grassy Mountain, Utah). Garbage Project records suggest that some 350,000 bottles of nail polish are thrown out in Tucson every year. The bottles that have been recovered intact have usually contained significant amounts of polish; most bottles of nail polish in landfills get broken.

Nail polish in landfills does not in itself constitute a serious or frightening problem, but it demonstrates that the means by which hazardous substances travel from household to place of final disposal are as manifold as they are mundane. And the amount of hazardous waste in municipal landfills is not negligible: According to Garbage Project studies, about one percent by weight of all garbage coming directly from households can be regarded as hazardous (based on EPA definitions of that term). That is quite enough to cause trouble, particularly if a dumping site has been poorly located. In 1972, for example, scientists from the Environmental Protection Agency conducted a study of a dump in Norman, Oklahoma, that had been placed on highly permeable ground in an area where the water table was high. They discovered that a variety of industrial organic chemicals—ethyl carbonate, triethyl phosphate, dicyclohexyl phthalate, and a number of other challenges to orthographic serenity—had begun to leach slowly into nearby ground water. Further investigation revealed, however, that the source of these chemicals was not industrial waste: In its lifetime, the Norman dump had never received any. The source must therefore be common commercial and household products. Ethyl carbonate is a stabilizer and cosolvent for pesticides, fumigants, and cosmetics. Triethyl phosphate is a plasticizer for resins, plastics, pesticides, solvents, and lacquer remover. Dicyclohexyl phthalate is a plasticizer for rubber and polyvinyl chloride (the plastic used for packaging most household cleaners and shampoos).

Over the years the Garbage Project has undertaken several investigations of hazardous waste in garbage, ranging from an assessment of the effectiveness of Marin County's "Toxics Away!" Day, discussed in chapter three, to an analysis of how (and how promiscuously) toxic substances in landfills migrate from one place to another. Some of the investigations have involved garbage from landfills, and others have involved what is known as a hazardous waste "pull" using garbage collected directly from household trash cans. In the mid-1980s Douglas Wilson drew up a sheet of instructions for Garbage Project personnel involved in hazardous-waste pulls, and a small portion of it—a portion meant to offer a hypothetical example of items of hazardous waste being identified and recorded with various codes—gives some idea of the fastidiousness of the procedures involved:

EXAMPLE

D. Wilson and W. Hughes are doing a Hazardous Waste Pull. Wilson records Wilson/Hughes in the "Recorders" space and *9/1/85* (the present date—month, day, year) in the "Date of Analysis" line. Hughes weighs two bags which are the household refuse put out by household *A* in census tract *19* for collection on *8/31/85*. Wilson records the total weight (*31.5* lbs.) in the appropriate columns and the census tract and date of collection. The two separate out 4 pesticide containers and 2 automotive motor oil containers from the refuse.

ITEM

A. 1 22 oz. *Blackflag* ant and roach aerosol spray
B. 1 22 oz. *Blackflag* "roach motel"
C. 1 22 oz. *Raid* ant and roach aerosol spray with ⅔ waste
D. 1 22 oz. *Raid* ant and roach aerosol spray
E. 1 32 oz. *Pennzoil* motor oil
F. 1 32 oz. *Raylube* motor oil

The two separate the oil from the pesticides. They subdivide the pesticides into three groups: (A), (B), and (C,D). They subdivide the motor oil into two groups: (E) and (F). Wilson records the two *Raid* containers as follows: 041, 2, 44 oz., X (waste—on margin "⅔ :235 g"), RAID, ANTANDRO, J [code letter for aerosol can]. The other groups are recorded in similar fashion. Item C is saved for proper disposal and the rest, which have no contents, are discarded into the dumpster.

Tedious as data collection of this kind may be, it is an essential component of garbage analysis. And such data, in conjunction with other tools, including surveys, have led to a variety of insights into hazardous waste. To begin with, comparisons of the relative volume of toxic materials in today's Sunnyvale landfill and in a dump in

Florence, Arizona, that was active between 1900 and 1945 confirm what most people would suspect: Contemporary society is dependent on far more products with a toxic potential than was the society inhabited by our grandparents and great-grandparents. The Florence excavations reveal, however, that by the 1940s the quantity of hazardous waste entering the dump there had risen to "modern" levels (at least for the types of hazardous waste studied, mainly heavy metals). The results suggest that Americans have been producers of highly contaminated garbage not for ten or twenty years but for at least half a century.

And yet, as another Garbage Project study shows, people generally have no idea of the amount—or the nature—of the hazardous waste they discard. When asked, residents of Marin County tended to say that most of the hazardous waste they throw away consists of motor oil (46.1 percent of household hazardous waste discards, according to self-reports) and paints and thinners (41.4 percent). Their actual garbage says otherwise: People throw out a lot less paint (28.1 percent) and motor oil (23.0 percent) than they think, and a lot more household cleaners and pesticides (37.5 percent) than they suspect.

To what extent are households, as opposed to local businesses, responsible for the hazardous waste that winds up in municipal landfills? In the search for villains the average person would probably find it inconceivable that a town's residences throw out as much in the way of toxic substances as all of the town's "Small Quantity Generators"—its restaurants, retailers, and service companies; its gas stations, dry-cleaning establishments, and shops for developing photographs. But in fact, according to a Garbage Project study based on discard patterns in Marin County, New Orleans, Phoenix, and Tucson, residences are doing just that. (It is not true, however, that municipal landfills, in terms of toxicity, are indistinguishable from industrial waste landfills—a claim that one occasionally hears, particularly from some spokesmen for manufacturing groups.)

The bad news in all this is that ordinary municipal landfills contain far more hazardous waste than most people tend to assume, even when such landfills have not been subject to the dumping (legal or illegal) of industrial chemicals. But Garbage Project investigations have unearthed some good news as well: In the landfills that have been studied, many kinds of toxic waste pretty much stay put. As we will see, even if these toxic substances ooze from a broken jar or leach out of paper or metal, they tend to be absorbed by the surrounding matrix material, and hunker down for a long stay.

One substance to which the Garbage Project has paid considerable attention is lead, both because lead poisoning is a source of increasing concern in the United States (the Department of Health and Human Services in 1985 lowered the threshold for what is considered lead poisoning from a level of 30 micrograms per deciliter of blood to 25 micrograms, and it lowered the threshold again—to 10 micrograms—last October) and because lead has played such a problematic role in the course of human history. Lead also nicely serves to demonstrate the unexpected ways in which a toxic substance can get into a landfill—and what happens to it when it does.

During the past three decades, archaeologists have focused vigorously on the migration of various kinds of hazardous substances out of such things as pots and pans and pipes, and into the people who used them. Most of the archaeological investigations that have been done, as Douglas Wilson observed in a recent study, have looked into the possible effects of lead—one of the most toxic of heavy metals but also one of the most widely used, owing to its malleability, its low melting point, and its resistance to corrosion. For thousands of years lead was used as a flux to lower the melting point of ceramic glazes in order to produce the desired glass-like coating. In the Roman Empire, lead was commonly employed in the manufacture of pipes for drinking water, as a lining for vessels made of bronze, and in paint. Various analyses of bones exhumed from ancient Roman cemeteries in Great Britain indicate that an elevated level of lead in the body was far more prevalent during the days of the Roman Empire than it is today. (In 1965 the historian S. Culum Gilfillen suggested in an article in *Mankind Quarterly* that brain damage resulting from widespread and sustained lead poisoning was an important contributor to the decline of Roman civilization—a view that, however, is firmly if gently dismissed by most other historians.)

Exhumations of eighteenth-century skeletons at one old Virginia plantation show that the Dominion's aristocrats may have paid a price in terms of health for using costly, lead-glazed porcelain dishes, such as those made by Josiah Wedgwood, and pewter vessels for storing food and drink. Owing to such conspicuous consumption, the lords of the manor and the members of their families had levels of lead in their bones five times higher than that of the slaves and laborers buried nearby.

Lead found a new vehicle for bodily invasion in the form of the tin-can food container, which was patented in England in 1811, and was immediately put to wide use by the far-flung Royal Navy. These containers had lead solder down the main seam and also lead solder plugging the sealing hole on the top. Some insight into the impact of a diet that was high in tinned food during the nineteenth century was provided by a Canadian physical anthropologist, Owen Beattie, and an Arctic archaeologist, James Savelle, who in 1981 launched a fresh inquiry into the causes of death of the members of the Franklin Expedition, which had set out in 1846 to discover the Northwest Passage. In 1848 the expedition's crew abandoned ship after their vessels, the *Erebus* and *Terror*, were caught in Arctic ice. No one survived. The Beattie-Savelle team discovered the frozen bodies of two of the British sailors, perfectly preserved, and removed pieces of tissue for tests. The consequences of the explorers' shipboard subsistence on tinned foods was pronounced in terms of the presence of lead, which had built up to catastrophic levels. The lead would have weakened the men severely and, Beattie and Savelle speculated, greatly impaired their judgment.

Despite everything that we now know about lead, and about its potential consequences if ingested, lead still turns up in landfilled garbage in alarming quantities. In 1988, as part of an experiment by Douglas Wilson, eighteen cans were randomly pulled from garbage buried in 1976 at the Sunnyvale landfill. All eighteen had a lead-solder seam on the side. Each of them also showed signs of rust (the one degradation process that occurs very quickly in landfills), suggesting an overall breakdown that could result in the release of the lead into the adjacent sediment, or "fines." Garbage Project records of all the sorted pre-1988 garbage from Sunnyvale indicate that 258 lead solder–seam cans were recovered. Lead is currently being deposited in landfills by many sources other than cans. Motor oil that contains lead accounts for about 0.02 percent of all residential refuse

—a relatively small amount, though in the city of Tucson, still equivalent to some twenty-two tons of motor oil a year. There is of course a considerable amount of lead in lead-acid car batteries. Light bulbs almost always have a bit of lead solder on their metallic bases. Another contributor of lead to landfills are the inks in old newspapers and magazines. Newspapers have taken big steps during the past decade and a half to drastically reduce the amount of lead in the inks they use, but this, of course, can have no effect on the many decades worth of newspapers that have already gone to their final resting place.

What happens to lead and other hazardous substances after they are disposed of? Do they begin to separate from their carriers and sink deep into the landfill with the first heavy rain that breaches the dirt cap? Research by the Garbage Project suggests that concentrations of heavy metals, to consider just one type of toxic waste, do migrate away from their point of entry into a landfill—but, fortunately, not very far, usually attaching themselves to particles in the immediate vicinity. Garbage Project researchers came to this conclusion after a series of tests conducted in 1990 on samples of fines from five landfills: twenty-four samples from Fresh Kills, twenty-three from Collier County, eighteen from Sunnyvale, seventeen from Rio Salado, and ten from Naples Airport. Four control samples were collected from untainted soils in the vicinity of Rio Salado. Portions of all of the samples were dried and sieved and finally subjected to absorption assays to identify concentrations of arsenic, mercury, lead, zinc, and cadmium.

There isn't much arsenic in municipal solid waste, the assays revealed: The levels in the landfill samples and the control samples were a virtual match. But the level of mercury was clearly higher in the landfill than in the control samples—perhaps owing to the presence of mercury in floodlights and fluorescent lamps, and in the "button" batteries found in every small appliance from watches to calculators to hand-held Nintendo games. Mercury appears to be highly separable. (Once used in the treatment of felt for hats, it can cause poisoning that results in neurological dysfunction; hence the phrase "mad as a hatter.") In many but not all of the landfill samples the concentrations of lead, cadmium (which is found in certain kinds of batteries, in PVC plastics, and in many paints and inks), and zinc (which can be found in dry-cell batteries and in lubricating oil) were substantially higher than they were in the control samples. The fact that heavy metals in landfills do migrate into the fines seems hard to argue with.

The question of whether the metals migrate very far was approached in another way. The assumption was made, first, that the natural direction of migration in a landfill would be from the upper reaches toward the lower reaches, and, second, that migration could be assumed to be great if lower (older) samples showed significantly larger concentrations of heavy metals than higher (younger) ones with similar contents. Those assumptions made, the concentration of each heavy metal in each sample was correlated with the length of time the sample had been buried, the depth from which the sample had been extracted, and the moisture content of the fines. When the data were analyzed, no clear-cut patterns emerged, suggesting that whatever migration occurs in a landfill is modest and localized. As the Garbage Project's Masakazu Tani observed in one study of excavated landfills, "It is unlikely that any substantial amount of heavy metals moves downward, presumably because any leachate flow is either blocked, absorbed, or filtered by the massive amount of dirt and refuse."

One piece of data dramatically supported that point. A sample with one of the lowest concentrations of lead (350 ppm) of all the samples examined was Sunnyvale sample 5-3 (from the 1964 stratum of Sunnyvale's well 5). For twenty-eight years the fines in this sample lay directly below sample 5-2, which contained one of the highest lead concentrations (2,500 ppm). Sample 5-2 was from a bucketload taken at a depth of fifty-five feet. Sample 5-3 was the very next bucketload. The heavy lead concentrations in the soil of the top sample had simply failed, for whatever reason, to filter down into the soil of the sample a few inches below.

Throughout history nature has from time to time contrived to pass on the manifestations of *Homo sapiens* from one generation to another. In the high, dry mountain valleys of Chile and Bolivia and Peru bodies have been preserved in a mummified state for many centuries, sitting, still dressed in the clothing in which loved ones had placed them. Not only bodies but artifacts of every kind that ordinarily should have fallen victim to the ravages of air and water and age have instead, through one fluke or another, crossed the centuries and been found at our doorstep. The flukes in question are accidents sometimes of climate, sometimes of geography, sometimes of chemistry. At the Town Creek Indian Mound site in North Carolina, the skeletal remains of a human being from the Late Mississippian Period (c. A.D. 1350) were discovered in 1952, and on the person's shroud there had been laid a copper amulet in the shape of a lizard, and, because copper oxide acts as a powerful preservative of organic matter, a piece of the man's shroud in the shape of a lizard was discovered under the amulet—the only fragment of the shroud that survived. A more startling example, from 1991, was the discovery of the almost perfectly preserved body, packed in glacial ice in the Austrian Alps, of a man who died some 4,000 years ago.

Archaeologists have depended these many years on chance's bounty, and will for many years to come. But thanks to their sanitary landfills the inhabitants of the twentieth century are doing Nature an order of magnitude better. For these Monstrous Visual Symbols are also time capsules, and if what we believe we know about biodegradation is correct, they will be durable beyond precedent.

PART III

Interlude: Diapers and Demographics

CHAPTER 6

A GARBAGE CENSUS?

If you were to identify a hundred households that were home to at least one cat but not to any dogs, here is something that you would learn from garbage about those households: If you collected their garbage for five weeks, you would find that at some point during that period 30 percent of the households had thrown away a copy of *The National Enquirer*.

If you were to identify a group of households that were home to at least one cat and at least one dog, data derived from garbage could tell you this: Cats get a better quality of cat food than the dog or dogs in the same household get of dog food.

If you were to look at garbage from a group of Hispanic households, here is one thing you would discover: The most popular baby-food vegetable by far is squash, which accounts for some 38 percent of the baby-food vegetables that Hispanics consume (squash has been a dietary staple in Mexico and Central America for at least nine thousand years); among Anglos, in contrast, the most popular baby-food vegetable is peas (accounting for 29 percent of all baby-food vegetables consumed), with squash ranking next to last in terms of preference, just above spinach.

If you were to look at household garbage for food waste from fast-food hamburger take-out restaurants and fast-food chicken take-out restaurants, you would learn the following: The food waste from the chicken restaurants (about 35 percent of all food bought, by weight, not counting bones) is considerably greater than the food waste from the hamburger ones (only about 7 percent of all food bought).

If you were to look at the garbage thrown away by the people in Mexico City you would learn this: that the urban poor consume proportionately more candy—pound for pound, the most expensive type of food in the city—than their more affluent neighbors.

All of these findings reflect phenomena that are regular and reliable—phenomena that are durable at least in part because they involve the behavior, as reflected in garbage, not of one person or a mere handful but of large numbers of people. The odd family may be located that eats everything it brought home from a fast-food chicken restaurant, or treats its dog better than its cat, but such peculiarities are not sufficient to mask our picture of mightier tendencies. The patterns of the various kinds of behavior that link masses of people and masses of garbage are, in the aggregate, strong and stable—so stable, indeed, that one can predict how much of what sorts of garbage a group of a given size and given nature will produce. Garbage thus becomes an important demographic tool, a point that was touched upon in chapter three with respect to certain unwitting consumption and discard habits, and the quirks of mind that lie behind them. But the applications are broader still.

When confronted on one occasion with the question of whether the physical laws of the universe were in any way irregular or capricious, Albert Einstein observed: "I shall never believe that God plays dice with the world." The fact that He does not is what makes the natural sciences possible. Garbology is made possible by the fact that garbage, too, is not a game of chance, at least when considered in quantities larger than those generated by a household or two. The accumulation and disposal of garbage is governed by what can almost be seen as natural laws (such as the First Principle of Food Waste). Garbage is, in a way, part of an ecosystem, one whose properties depend on the weather, the season of the year, and the rate of growth (or decline) of the Gross National Product. Perhaps not surprisingly, the solid-waste stream becomes a little thinner during times of recession than it is when the economy is robust (with a marked fall-off in construction and demolition debris, in particular). Municipal solid waste gets wetter and therefore heavier in the summertime (a fact that has bedeviled some incinerators, which yearn for dryness and above all consistency in the material they must process), and summer is also the time when garbage is most thickly studded with valuable aluminum beer and soda cans (but also the time when it contains the lowest percentage of paper and plastic, thanks to the enormous warm-weather influx of yard waste and construction and demolition debris). The properties of garbage also vary according to the presence of various types of human fauna: That is to say, they vary markedly—but also predictably—from one kind of neighborhood to another. A trash bag containing many regular soda cans and relatively little discarded food has a high probability of having originated in a Hispanic household. Garbage with lots of packaging from "status" brand-name foods and drinks is more likely to have come from a middle-income neighborhood than from an affluent one, while an affluent neighborhood is most likely to discard containers that once held diet soft drinks and store-brand and generic foods.

The obvious question is: What's the point of knowing all this—of having this demographic tool? There are two answers, one practical and one theoretical. The practical answer is that demographic data derived unobtrusively from garbage have a variety of real-world uses—in marketing and consumer research; in the rational governance of communities; in any endeavor that demands detailed knowledge of the behavior (including the relatively private behavior) of large groups of real people. Let us say that a supermarket chain in a certain city has advertised a special discount price on a specific kind of detergent. Garbage sampling over a period of weeks could reveal: a) the percentage of households that proved susceptible to the advertisement; b) the percentage of the households buying the detergent in which it was put to use immediately rather than stockpiled; c) the percentage of households that bought the detergent and then bought it a second time; and d) the percentage of households that bought the detergent, didn't like the results, and threw the rest away.

In a 1984 study published in a special issue of the journal *American Behavioral Scientist* that was devoted exclusively to garbology, Michael D. Reilly, a professor of management and marketing at Montana State University, set out to illustrate some of the possibilities inherent in garbage for market research. In one instance, Reilly showed how data from garbage could be used to explore the strength of brand loyalty among users of different brands of any type of product. One can do this, Reilly explained, by looking at how many different brands of a specific type of product are to be found in garbage from a household where a particular brand of the product predominates. Directing his attention to cat food, Reilly discovered that garbage samples from households that tended predominately to use 9 Lives cat food also contained a lot of cans from other brands of cat food—considerably more "disloyal" cans than did garbage samples from households that tended to rely predominately on a brand of cat food other than 9 Lives. In other words, Reilly concluded, the brand loyalty of 9 Lives users is relatively weak. (Reilly's calculations, which are based on Garbage Project data, are also the source for the marketing information about fast-food restaurants and cat and dog owners cited at the outset of this chapter.)

One legendary example of the practical applications of demographic garbology derives from the 1930s, when *The Saturday Evening Post,* which had a large middle-class readership, sought to convince the Campbell Soup Company—as yet not an advertiser in the *Post*'s pages—that the households of average businessmen and their families, not those of the truly affluent, were the chief buyers of canned soup. To counter Campbell's claim that less-affluent households probably made their soup from scratch, the *Post* collected large amounts of garbage from middle-class neighborhoods in Philadelphia and also from neighborhoods in Philadelphia's prosperous Main Line, dumped them on the floor of a local armory, and proceeded to look for soup cans. The results were precisely the opposite of what Campbell had assumed. Rich households were much less likely than middle-income ones to use canned soup—they employed servants, who could prepare soup from fresh ingredients. Middle-class households, in contrast, enjoyed little leisure time and had little domestic help. To judge from their garbage, they appreciated the convenience of canned soup.

The theoretical answer to the question "What's the point?" is that garbage offers a way to investigate issues of "mentality." As noted earlier, there are currents in the mind of populations and subpopulations that are sometimes difficult to glimpse—indeed, that may perhaps never have been apprehended. Here is a modest example. In 1986 the Garbage Project was asked by Heinz, U.S.A., to undertake a study in Tucson of various aspects of baby-food consumption in the United States. One odd phenomenon that was identified accidentally—Garbage Project researchers were not looking for it or, for that matter, even aware of its existence—was the desire of some of the Hispanic women surveyed to leave the impression that they prepared all of their baby food from fresh materials. In the course of the study, members of the households whose garbage was going to be analyzed had been asked the following question (in English or Spanish): "In the past seven days did your household use any commercially prepared baby food or junior foods?" Not a single Hispanic mother admitted to using even one jar of baby food; the garbage from the Hispanic households nevertheless contained just as many baby-food jars as did garbage from other households with infants. The Hispanic mothers, it seems clear, were reporting adherence to an idealized form of cultural behavior that (in a community where 45 percent of all Hispanic women are in the labor force, and therefore often do not have time to prepare traditional meals) may not have been the norm for many years.

The Garbage Project started out in the early 1970s as an archaeological experiment centered on material culture and "formation processes"; saw its attention focused heavily throughout its first decade on a variety of issues having to do with food and with recycling; and eventually concentrated much of its time and effort on the understanding of landfills, garbage-disposal technology, and the nature of the garbage crisis. In all of these endeavors, however, demographic issues of both a practical and theoretical kind have never been far below the surface (if they were below the surface at all). And one important series of studies, conducted on behalf of the U.S. Bureau of the Census, was as focused on demographic matters as can possibly be.

In 1986 the Census Bureau came to the Garbage Project and explained that it had what is by now a much-publicized problem: the problem of undercounting. Most of those people whom one can characterize as inhabiting mainstream America do get counted by Census Bureau tabulators; these people either return their census forms promptly or, failing that, are visited by Census Bureau personnel. There have always been difficulties caused by the fact that Americans are mobile, their addresses therefore changing with some frequency, and by the fact that some Americans are ornery, and just don't want to cooperate with the government. But the biggest challenge for the Census Bureau in recent years has been caused by the millions of people who live in society's nether world: undocumented aliens, the homeless, and many of the residents of America's urban ghettos (especially men)—people who are poor, in many cases illiterate, and perhaps fearful of the census (if aware of it at all), despite the government's promise of confidentiality.

A census undercount is a matter of concern for many reasons. The reason one hears about most often is financial—some $35 billion annually is transferred from Washington to the fifty states on the basis of the Census Bureau's population data—but the census also has a major impact on legislative redistricting and in the way government programs are administered and on how social scientists make sense of what is happening inside the country. And while complaints about undercounting are not new—Thomas Jefferson, the director of the 1790 census, was not happy with the final tally of 3.9 million in that year, believing it to be low by about a hundred thousand—the growing extent of the undercount has attracted critics both inside and outside the government. It is believed that the 1990 census undercounted the U.S. population by as many as six million people, which would be the largest undercount in American history. The Census Bureau, it should be noted, had long been prepared for this eventuality, and it has developed statistical procedures for correcting the undercount even down to the local level. The reason the Census Bureau came to the Garbage Project was to see whether data derived from garbage could reliably be used to check the Bureau's survey data in problematic neighborhoods, particularly neighborhoods where there was reason to believe the census was missing large numbers of adult minority men.

Missing minority men have long been a big problem. One study conducted in 1968–1971 by the cultural anthropologists Charles A. and Betty Lou Valentine, who lived for a long period as "participant-observers" in a public-housing-project highrise in a Philadelphia ghetto, found that many of the men whom they knew to be members of households in the apartment building had escaped the Census Bureau's notice—the undercount of men, by their estimate, was 61 percent—largely out of fear that welfare, immigration, or law-enforcement authorities would ultimately have access to the Bureau's questionnaires. "The local surveys," the Valentines wrote of the Census Bureau effort in the housing project, "make it appear that over 70 percent of the adult population are women, but ethnographic observation produced a sex ratio very close to 1:1."

With difficulties such as this in mind it was decided that the Garbage Project would look at some of the garbage data it already had on hand to see if it was possible to roughly reconstruct a community's population by age and sex simply on the basis of what that community threw away. The assignment was given to the Garbage Project's Masakazu Tani, who had come to the University of Arizona from Japan in 1981 to pursue graduate studies in archaeology, and who, as a foreigner, might display a stance toward living American culture that was in many ways analogous to that displayed by an archaeologist toward any ancient dead one.

To an archaeologist, the idea of using garbage to reconstruct population characteristics is neither bizarre nor uncommon. Archaeologists are frequently faced with the problem of needing to know the size of some particular past community, and having no written records to go on. Indeed, this is one of the classic problems of archaeology, and many techniques for arriving at population size have been tried. They can be fairly blunt instruments. "Archaeological population estimates," says Jeffrey Reid, an archaeologist at the University of Arizona and the editor of the prestigious archaeological journal *American Antiquity*, "are just the number of identified dwellings multiplied by some magic number that is a guess at an average number of household residents."

The magic number can be arrived at in a variety of ways. One is by cross-cultural analogy. The archaeologists Sherburne Cook and Robert Heizer, for example, used ethnological studies and other first-hand accounts of living Native American communities to try to determine how many people typically inhabit how much roofed space; the idea, of course, was that the latter-day dwellings of aboriginal people could serve as a stand-in for ancient ones. Cook and Heizer's search for a universally valid correspondence between household size and roofed space—a classic study in the field—foundered somewhat on the fact of climatic differences: People in hot climes require much less indoor space than people in cold ones. Still, Cook would later write, a serviceable rule of thumb for relating ancient floor space to number of inhabitants "is to count 25 square feet for each of the first six persons and then 100 square feet for each additional individual."

Even if the methodology is not cross-cultural, it necessarily involves defining a numerical relationship between archaeological remains (huts, fire pits, broken pottery, burials, the size of a site as a whole) and the individuals responsible for them. Some of the techniques are ingenious, as, for example, in the case of a famous 1966 study by Christy G. Turner II and Laurel Lofgren, in which the archaeologists attempted to establish the sizes of prehistoric Anasazi Indian households by dividing the volume of the average serving bowl into the volume of the average cooking pot; this calculation, when done with many bowls and pots from several different periods of time, would, Turner and Lofgren surmised, bring broad demographic trends into relief. (The assumption, of course, was that a rise or fall in the ratio of cooking-pot size to serving-bowl size must reflect a rise or fall in the number of people to be served per household.) In the end, Turner and Lofgren estimated that between A.D. 300 and A.D. 1300, the average Anasazi household size increased from 4.5 members to 5.2 members.

Many archaeologists are quite conservative and report on population only in relative terms. If the potsherds from period A seem to be scattered over a wider area than the potsherds from period B, then the population of A is usually assumed to have been larger than that of B. However, this approach can run into trouble, too, it turns out: The density and distribution of "scatter" can relate not only to the number of people inhabiting an area but also to the length of time a group of people inhabits an area. Recall, for example, the tendency of Australian aborigines to move on when the volume of nearby garbage reaches a certain threshold.

In effect, the Garbage Project's Census Bureau study was as much a search for a "magic number"—a multiplier that, when applied to certain quantities of certain kinds of garbage, would yield accurate population estimates—as any archaeological investigation in the field. But the Garbage Project started out with advantages that other archaeologists could only envy. An archaeologist in the field doesn't know what percentage of all the artifacts once present is represented by the discards he finds or digs up in and around the remains of a dwelling. He also doesn't necessarily have a clear sense of whether the artifacts he finds were all deposited in the places where he found them at roughly the same period of time or over a relatively long period of time. Even more to the point, an archaeologist has no way to check his population methodology against an actual body count.

The Garbage Project, in contrast, had, among other information, the mass of computerized evidence compiled during its studies of food consumption for the Department of Agriculture: data that it knew to represent a complete inventory of household garbage from sixty-three separate dwellings, and that it knew to have been collected over a specific five-week period. Because the Project staff had also conducted personal interviews with members of all of those sixty-three households, it had a complete record of exactly how many people inhabited each one, and it knew the sex and age of each person. The Project interviewers, it should be noted, were especially vigilant against possible misreports by household members as to who actually ate and slept in respondent households. For the purpose of establishing a correspondence between elements of material culture—in this case, garbage—and size and composition of population, the situation was nearly ideal.

The first step was to come up with a way of making the leap from the quantity of garbage, or the quantity of a particular type of garbage, to an overall population estimate. One could begin, of course, with the knowledge—confirmed by various Garbage Project studies—that in the United States, in general, the larger the household the more garbage it tends to produce. The idea now was to come up with a number which, if you multiplied it by the weight of all the garbage—or that of all of one of sixteen particular subtypes of garbage—collected from a certain group of households over a specified period of time, would yield a number for the population inhabiting those households. The categories into which garbage was broken down for the purpose of the population-estimation experiment were the following:

Total solid waste	Ceramics
Packaging paper	Plastic
Non-packaging paper	Ferrous metals
Newspaper	Bi-metal
Magazines	Other metal
Textiles	Aluminum
Food debris	Nonreturnable glass
Yard waste	Returnable glass
	Other

Briefly, the weight data for all of these categories was plotted on graphs against data on household size, dwelling by dwelling, and the resulting "scatter plots" were analyzed to see in which categories the weight data showed a steady, monotonic rise relative to household size. There's probably no point getting into much more detail than that: A lot of not very interesting mathematical manipulations were simply made to occur, and the result was that the categories "Total solid waste" and "Plastic" were found to have the most accurate predictive power. It was on the basis of these that the correlation equations were derived. The equation derived from total solid waste is, however, less universally reliable than the one for plastic; this is because children are responsible for less garbage overall than are adults (roughly speaking, one adult's garbage equals that of 1.45 children), and so population counts would be skewed by data from a neighborhood whose ratio of adults to children was markedly different from that in the sixty-three original sample households. Plastic is another story, however: During any given period of time, every man, woman, and child in America generates about the same amount of plastic garbage, usually in the form of many, many, small items (and accounting for about 0.52 pounds of the 7.40 pounds of garbage that is thrown away in a week by a typical individual). Plastic is America's great garbage equalizer. The following equation, which has been stripped of certain technical apparatus, expresses the rough relationship between plastic and population:

$$\text{Number of People} = 0.2815 \times \text{PLASTIC}$$

where the PLASTIC quantity is based
on a five-week collection and recorded in pounds

You probably won't want to attempt to validate this equation using your own household—as noted, the amount of garbage required is five weeks' worth, and even though only the plastic need be saved (and not, say, the slops), the amount of sorting and separating and cleaning involved would be considerable. Also, while the equation is reasonably accurate at the household level, it is really designed for use with respect to whole neighborhoods, where the biases and "noise" that could easily skew a household-population figure up or down by one person are evened out by the sheer volume of a neighborhood's garbage. For a neighborhood of some 100 households, the projected total population estimate derived from Garbage Project equations applied to five weeks' worth of garbage will be accurate to within plus or minus 2.5 percent. That is considerably better than the Census Bureau can do in many places—and, indeed, better than the Census Bureau actually did for some groups. The undercount of black men in 1990 has been optimistically guessed by the Census Bureau to be some 5.9 percent, and the undercount of Hispanic men is estimated to have been 6.2 percent. The undercount for the entire District of Columbia was some 7 percent, an average that hides considerably larger undercounts in many neighborhoods.

Overall population is only one demographic characteristic, however. What about producing estimates by age and sex? This turned out to be—as Project staff members suspected it would—a trickier proposition. The easiest subpopulation to discern is infants. Disposable diapers are a convenient marker for this group of people, and infants go through so many diapers that they are an ideal item for establishing correlations: the regularity and volume of their disposal in household garbage helps suppress statistical biases. Here's the rough equation for estimating the number of babies in any given neighborhood:

$$\text{Infant Population} = 0.01506 \times (\text{Number of diapers})$$

where the number of diapers is based on a
five-week collection

Any parent dividing 1.0 by 0.01506—which yields the average number of diapers collected per infant in a five-week period—will be surprised at how low the number is. It must be remembered that the equation involves the number of diapers recorded in *household* garbage. That figure is not the same as the number of diapers an infant uses. Diapers get thrown away at daycare centers, at malls and supermarkets, in sidewalk trash cans. The fact that a significant fraction of parents use cloth diapers further depresses any neighborhood's household average.

Unfortunately, there is no one item of garbage that can help us determine the proportion of men and women in a population, or the proportion of children and old people—none, at any rate, with the ease and power that disposable diapers display with respect to infants. It is not that men and women (or children and old people) do not leave traces. Anybody given four or five bags of garbage and told that they came from a household of five could probably puzzle over it all—puzzle over, say, the gnarled tube of Dentucreme, the nylon stockings, the pages from a Barbie coloring book, the empty Amphora pipe-tobacco pouch, the baseball-card wrapper, and the many other suggestive products—and come up with a fairly good guess as to the household's demographic configuration. It is quite another matter, however, to establish that configuration for a whole community from a mass of undifferentiated garbage.

Making the job even harder is that the material correlates for various demographic groups tend to become a little unreliable as one leaves infancy behind. As a demographic marker, for example, disposable diapers for infants are what is known as an "exclusive": only infants wear them. To be sure, one may find a few used diapers in the garbage of a person, such as a grandparent or babysitter, with no infant in permanent residence. But the diapers themselves point to the existence of babies.*

When it comes, however, to distinguishing between men and women, or middle-aged adults and elderly ones, there are fewer exclusives and far more "proportionals"—items that, like disposable razors, may *tend* to indicate the presence of members of a particular demographic group (in the case of razors, adult men), but don't necessarily do so by any means. Moreover, the exclusives or near-exclusives that do exist for these groups are primarily items that (again, unlike diapers) are discarded very infrequently and thus, for demographic purposes, have little predictive value. In the case of the elderly, for example, a copy of *Modern Maturity*, the magazine of the American Association of Retired Persons, may stay on the living-room coffee table for six weeks before being discarded (if it ever is); telltale prescription drugs often don't need to be refilled for months, and the empty containers sometimes stay in the medicine cabinet.

It took a considerable amount of work, but it eventually proved possible to come up with an equation for estimating the proportion of children in a population, based on the average number of discarded toys and toy packages (2.52 per child per week) and the average number of discarded children's clothes and clothes packaging (0.87 per child per week). It was also possible to derive equations for estimating the proportion of adult women in a population, based on the number of discarded female-hygiene packaging (1.58 per woman per week), cosmetics (0.86 per woman per week), and clothing items (0.62 per woman per week).

Finding serviceable material correlates for estimating the proportion of adult men—the Census Bureau's ultimate objective—proved more elusive. Men are not exactly invisible in garbage, but garbage is a more unreliable indicator of their live-in presence than it is for any other demographic group. Women may drink and eat like men. They smoke cigarettes. They sometimes wear men's clothing and

* Or so one would like to presume. It is conceivable, of course, that there is an equivalent for disposable diapers of the cloth-diaper fetishist in St. Petersburg, Florida, who posed as a diaper-service driver and stole diapers off people's porches. After the thief's eventual apprehension, in 1987—he was wearing a diaper under his clothes at the time—a search of his home turned up some 370 diapers, all cleaned and neatly folded.

men's cologne. Henry Higgins's plaintive query—"Why can't a woman be more like a man?"—is not one that passes the lips of a garbologist. Even the presence in garbage of male contraceptives or (more likely) contraceptive packaging is at best equivocal evidence of a male household member. While the wrappers and chewed butt-ends of fat cigars may be indicative of a live-in male, as may be the packaging from men's underwear, such diagnostic aids are exceedingly rare. Equations eventually were derived from underwear and a few other items for estimating the proportion of men, but these male-resident equations were afflicted by unacceptable margins of error.

In the end, the best way to get a figure for the number of adult males in a given neighborhood turns out to be a back-door procedure. First, find the total neighborhood population. Next, subtract the estimates for infants, children, and adult women from the total population estimate. The result is an estimate of the adult male population, and it has an accuracy of better than plus or minus 10 percent. Thus, if the actual population of adult males in a neighborhood was 240, the evidence from garbage would yield a range from 216 to 264.

Using garbage for population estimation would, of course, not be all that helpful in places like Greenwich or Shaker Heights, Aspen or Beverly Hills, where the Census Bureau's data are accurate down to the last scullery maid. But there is not much doubt that estimates derived from garbage could provide a usable snapshot of many neighborhoods that the government otherwise would find hard to penetrate. As it happened, however, the Garbage Project never got the chance. In 1988, the director of the Census Bureau's Center for Survey Methods Research decided that, from a public-relations standpoint, "It was risky for the government to hire someone to analyze garbage." A year later the Bureau announced a tentative decision not to adjust Census Bureau findings at all (by any method) in order to compensate for the expected undercount, a decision to which it has steadfastly adhered ever since.

Developing population models based on garbage demographics obviously depends, as noted, on the assumption that, in the aggregate, different types of people tend to throw away somewhat different types of garbage. The Census Bureau project was concerned with calculating the total population of a community and that community's demographic makeup by age and sex, with particular attention to the number of adult men. Garbage can also reveal differences by income and ethnicity, some of which were alluded to at the outset of this chapter.

But beyond establishing certain material correlates of the *fact* of affluence or poverty, or the *fact* of ethnicity—useful in their own right, to be sure—garbage can also offer some insight into issues of mentality. A case in point involves what has come to be known as the Hollywood Hypothesis.

The Hollywood Hypothesis, which was developed by Garbage Project personnel, can best be explained with reference to a study based on Project data that was conducted by Michael D. Reilly, who was mentioned above, and Melanie Wallendorf, who is an associate professor in the marketing department at the University of Arizona. Their idea was to test what has come to be seen as the "traditional" model of ethnic assimilation—namely, that "immigrants to a new culture will exhibit a cultural style that lies somewhere between the normatively prescribed behavior patterns prevalent in the culture of origin and those prevalent in the culture of residence"—by means of garbage analysis focused on food consumption.

Reilly and Wallendorf concentrated specifically on Mexican-Americans, and on their consumption behavior with respect to seven types of comestibles: meats and eggs, breads, cereals, coffee, soft drinks, alcohol, and convenience foods. With the help of the Garbage Project database, the researchers compiled numerical tables, corrected for household size, showing average daily household consumption in Anglo-American (Tucson), Mexican (Mexico City), and Mexican-American (Tucson) households of each of the seven food types, by weight or volume. The food groups were also broken down into subcategories (white bread and dark bread under the "breads" category, for example). It should be noted that Reilly and Wallendorf were of course aware from a variety of published studies that Anglo-American consumption patterns had been sharply affected by, and remained responsive to, several powerful trends: a shift away from red meat and toward poultry; a shift away from white and toward dark bread; a shift away from high-sugar cereals; a shift away from caffeine; a shift away from sugared soft drinks; a shift away from spirits and toward beer and wine; and frequent use of convenience foods. None of these trends had been much observed in Mexico.

The surprising conclusion of the study was that the habits of Mexican-Americans do not fall somewhere between those of Mexicans and those of Anglo-Americans—not in the least. It was to be expected that in virtually all the food categories the gross consumption figures for Anglo-Americans would always be higher than those for Mexicans, given the simple fact of greater affluence north of the border. But consumption figures for Mexican-Americans represent a further extreme. Take beef consumption: Where the average Anglo household consumes about 128 grams of beef a day, and the average Mexican household considerably less, the average Mexican-American household is off the charts, at 189 grams. (The Garbage Project initially logged a suspiciously low beef figure for Mexicans—2.9 grams of beef per household per day—is no doubt due in part to the fact that Mexicans buy much of their meat in open markets or small shops, and it therefore comes without labels that can find their way into the garbage. A comparison of the weight of discarded bones in the garbage of all three study populations showed that while Mexicans did eat more meat than the labels indicated, the phenomenon identified in the original findings was no less apparent.) Similarly, where the average Mexican household drinks 13.3 grams of sugar-based soda a day, and the average Anglo household drinks 169 grams, the average Mexican-American household drinks 291 grams. Mexican-Americans are also bigger drinkers of coffee and tea than members of the other two groups, heavier consumers of convenience foods, bigger eaters of eggs and high-sugar cereals, and bigger eaters of white bread. On the whole, there is little evidence that, insofar as food is concerned, Mexican-Americans have yet been much influenced by the currently evolving health-conscious Anglo life-style; it is certainly not the direction of adaptation. Rather, Mexican-Americans seem to have been assimilating toward the unreconstructed behavior patterns of Anglo-America as it was around 1965. "It may be," Reilly and Wallendorf write, "that Mexican-Americans have over-assimilated to their prior perceptions of Anglo cultural style." They go on: "This internalized conception of American life may trace back to representations encountered before migrating as well as to inferences drawn from the mass media and other depictions of American life encountered after the move." That, in a nutshell, is the Hollywood Hypothesis.

Reilly and the Garbage Project used the same methodology to look at lower- and upper-income households in both Mexico and the United States to determine whether appreciable differences existed between the two countries in terms of the way social status correlates with consumption. In this case the data derived from more than a thousand refuse samples from neighborhoods in Tucson, Milwaukee, and Marin County, on the one hand, and more than a thousand samples from neighborhoods in Mexico City, on the other. In some respects the patterns of consumption in both countries followed parallel tracks as one moved from lower- to higher-income households: Mexicans and Americans alike consumed more dairy products, more syrup and honey, and more liquor as their incomes rose.

But some odd divergences were apparent as well. Consider the case of canned vegetables. In the United States, a reliance on canned vegetables is a strictly low-end consumption pattern; canned vegetables are eaten by people who are not-so-well-off at twice the rate at which they are eaten by the most affluent consumers (who prefer their vegetables to be fresh, and perhaps even exotic). Among Mexicans, however, the most affluent consumers eat six times more canned vegetables per person than do the least-well-off consumers; they eat almost three times more canned vegetables than the most canned-vegetable-loving Americans do. The same pattern holds for cigarette consumption: In Mexico, it is the affluent, not the poor, who smoke the most cigarettes—the exact reverse of the situation in the United States. And the same pattern holds for toilet paper: Not only do affluent Mexicans use almost twice as much toilet paper per person (as suggested by a count of toilet paper rolls) as poor Mexicans, they use almost six times as much as affluent *Americans*. To be sure, in Mexico toilet paper often substitutes for other paper products. But why the disproportionate fondness among affluent Mexicans for canned vegetables? Why the disproportionate fondness for cigarettes? Is Mexico, in terms of consumption, following a path similar to one already taken by the United States, but lagging a few decades behind? Is it, in other words, chasing a moving target? Or is it in some ways heading in different directions altogether? For the moment, these must remain questions without answers.

Garbology in the service of demography is a novel discipline, and has as yet seen only modest applications. There will unquestionably be more, as social scientists and market researchers discover its value for collecting data without intruding into people's lives, while at the same time avoiding the biases that almost always creep into surveys and interviews.

And the Census Bureau? The Garbage Project is ready for the year 2000.

CHAPTER 7

THE DIAPER DILEMMA

Is there a more potent and pungent example of the quintessential character of the American solid-waste stream than the paper-and-plastic disposable diaper? *The New York Times* has named disposable diapers as the premier "symbol of the nation's garbage crisis." These diapers, which are known in the trade as single-use diapers, have a capacity to intrude upon our consciousness like few other kinds of garbage. We are aware of the bulging diaper pails at the homes of young friends, and of the vague aroma of sanitized plastic these young friends seem perpetually to carry around with them. We are aware of the disposable diapers we see atop garbage cans at zoos and parks, folded into tight, hermetic rolls and sealed with the remnants of adhesive on the already-used fasteners. We come across disposable diapers in the asphalt Gobi of once-crowded parking lots, where harried parents have changed their infants and then—ooops! —carelessly allowed the soiled diaper to slip to the ground through an ever-so-slightly opened car door. We see these diapers on the shoulders of highways and the banks of rivers, on picnic tables, on the incoming tide. We see them even if we try, briefly, to take a break from America by traveling abroad: Pampers, for example, are now available in more than eighty countries around the world. Robert W. Hollis captured something of the awesome international presence of the disposable diaper in a 1989 article in *Mothering* magazine. "I have personally seen excrement-filled diapers floating in the lagoons of Kwajalein and Majuro, two of the Marshall Islands in the western Pacific," he wrote. "When Ferdinand and Imelda Marcos arrived in Honolulu in exile from the Philippines, they were carrying jewelry, cash, and other booty in recycled Pampers boxes removed from Malacanang Palace."

Not only do disposable diapers compel our notice, they frequently do so in a way that calls attention to their bulk. In the home, they are usually not mixed in with all the other household waste; they are confined to a kind of isolation ward—a large, lumpy, *heavy* bag of their own, whose exterior bulges reflect the distinct identity of the diapers therein. Before the diapers can be used they must, of course, be bought at the store. In supermarkets across the country an entire side of an aisle is reserved for disposable diapers—as much as is taken up by all canned vegetables or soft drinks. Disposable diapers are among the very few items in supermarkets—jugs of cider and jumbo boxes of detergent are two others—that are so big or unwieldy that they never get bagged at the checkout counter, a fact that is taken into account in the design of packages for these products. Because they are so large, packages of disposable diapers are usually placed in an aisle that a methodical person will reach not at the beginning but toward the end of a shopping trip, just before bringing the cart about and setting course for the cash register. The large packages of diapers perched unsteadily on top of the groceries in shopping carts appear to be the equal in volume of everything underneath. The message is clear: The contribution of disposable diapers, rank and foul, to municipal solid waste must be enormous.

It is hardly surprising, then, that the use of disposable diapers by one couple will often raise the eyebrows of another, who would as soon throw beer cans out a car window in a national park as wrap their baby's bottom in anything but cloth. It is simply assumed by many people that *not* using disposable diapers is the environmentally correct thing to do, and this assumption seems to be shared by many environmental groups (more or less with a shrug, however; they haven't made disposable diapers a big public issue) and by many of those who do the reporting on garbage matters. A typical example of the kind of thing that gets written is this 1987 Associated Press story about Martha Gray, the designer of an exhibit about diapers at the Children's Museum in Holyoke, Massachusetts:

> Gray said she got the idea for the exhibit from a newspaper article about the environmental hazards of disposable diapers that end up in dumps.
>
> "I wondered, Could we be ignorant of the fact that we're creating a potential epidemic?" she said Monday. "That's raw, untreated waste and a potential carrier of disease. As consumers, we haven't thought about the impact of our decisions."
>
> To encourage parents to think about the hazards, she has made available articles on the subject, including one citing a 1978 study in Oregon that found 16 to 32 percent of the state's solid waste came from disposable diapers.

What is the truth about disposable diapers? Are the several thousand disposable diapers that the average child who wears them goes through during infancy—4,907, according to one recent study; 5,840, according to another—in fact less environmentally friendly than the cloth diapers that the same infant would have used, with all their washings and the generation of sewage and consumption of energy that washings entail? During the past decade and a half there have been a number of studies conducted that bear on the issue. Most of these studies were commissioned either by companies that manufacture disposable diapers, such as Procter & Gamble, or by the National Association of Diaper Service Industries, which for obvious reasons promotes the use of cloth diapers. The disposable-diaper and cloth-diaper forces are fundamentally at loggerheads, and we'll spell out their positions below. Fortunately, there have also been independent investigations of some aspects of the diaper issue, including some by the Garbage Project, and there are grounds for hope that at least a few elements of the diaper debate can be resolved.

The first recognizable example of a disposable diaper in the United States seems to have appeared in the years after the Second World

War, and was the invention of Johnson & Johnson. This diaper, called Chux, was a multi-ply cellulose product whose padding had the texture of paper towels. (A different sort of disposable diaper had been developed earlier, in Sweden. It consisted of disposable cellulose pads that were inserted into a reusable plastic cover, and was the kind of disposable diaper that some people propose we return to.) In 1961 the Procter & Gamble Company applied for a patent on a much more sophisticated disposable diaper to be called Pampers, which was heralded as "the discovery that makes diapers old-fashioned!" The development process had been somewhat rocky, and the diaper project was nearly canceled after a test of a two-piece version of the new product in Dallas in the late 1950s proved disastrous. This version called for a disposable pleated pad inside a plastic panty; the plastic panties made babies uncomfortable in the Dallas heat, and parents complained. It was back to the changing table. In fine corporate-realism style, *Eyes on Tomorrow,* an authorized history of Procter & Gamble, recreates the moment in 1958 when, after a brilliant presentation by Robert Duncan, the head of the diaper-research group, Gib Pleasants, the vice president for research, gave the go ahead for an innovative one-piece disposable diaper: "Pleasants looked across the desk and very quietly said, 'When you came in, Bob, I was of a mind to stop the project. I can't find it in my heart to stop it now. Test that new diaper.' " The test was successful. Today two companies, Procter & Gamble and Kimberley-Clark, produce some 80 percent of the more than 16 billion disposable diapers that Americans use every year. This figure does not include such brands of diapers as Attends and Depends, which are designed for incontinent adults; diapers for adults account for between 500 million and a billion of all the diapers Americans use in a year.

The first disposable diapers were somewhat different from the ones in use now; for one thing, parents had to use safety pins, rather than adhesive flaps, to keep the diapers fastened. Over the years the makers of disposable diapers introduced new fastening systems (including tape that can be *re*fastened) and elasticized leg cutouts for improved containment, and they light-weighted the volume of material needed for the manufacture of diapers and diaper packages by more than 50 percent. The "polybag" packaging now used for Pampers, which was introduced by Procter & Gamble in 1988, represents a nearly tenfold reduction in weight over the fiberboard packaging used earlier: from 0.018 pounds of packaging per diaper to 0.002 pounds per diaper, with a similar reduction in volume.

The typical disposable diaper is a many-faceted concoction. We tend to think of disposable diapers as being made of plastic, though in fact by weight only about 8 or 9 percent of a disposable diaper—the waterproof backsheet—is plastic. About three-fifths of a disposable diaper's constituent material by weight is plain cellulose, which goes into the diaper's absorbent padding; the padding is infused with a nontoxic polymer that turns into a gel when urine makes contact. The liner padding, fasteners, and a few other components together account for about another quarter of the diaper's weight.

The convenience of disposable diapers has proved to be so compelling an argument for their use that today they account for some 85 to 90 percent of all the diapering that is done in America. Convenience aside, disposable diapers seem to be more effective than cloth ones in the prevention of diaper rash, a conclusion drawn by, among others, *Consumer Reports* and the U.S. Office of Technology Assessment. Erma Bombeck probably spoke for the vast majority of America's parents in a recent column when she wrote: "As a mother, I'd rather do away with foam cups and have hot coffee poured into both of my hands and drink fast than do away with disposable diapers." All told, as one critic has pointed out, enough disposable diapers are thrown away in the United States every year "to stretch to the moon and back some seven times over." Even families who for perceived environmental reasons condemn disposable diapers are prone to use them when they find themselves traveling or must otherwise remove a child from the home environment; the diaper equivalent of "vegans"—vegetarians who are *really* vegetarians, forswearing even fish and dairy products—does not constitute a large bloc.

Once, on a television show, a Garbage Project researcher was asked to examine a week's worth of trash from the household of a member of the audience (with the family's permission, of course), the idea being to gauge the consistency of the household's level of environmental concern. He found many bottlecaps but no bottles, suggesting that returnables had been returned. He found paper labels but no metal food cans, suggesting that the cans had been flattened and recycled. He found food-preparation debris wrapped in single sheets of newspaper but no whole sections of newspaper, suggesting that newspapers had been bought and also recycled. The only significant amount of plastic he found consisted of some plastic containers for "natural" yogurt. In all this garbage one item stood out: a "giant, economy size" box of disposable diapers, which the mother in the household confessed were used to clothe her son at the daycare center (per the center's instructions). In practical terms, then, it may be that the war between disposable and cloth diapers has already been won (and lost).

This does not mean, however, that there is no room for a protracted, low-intensity guerrilla conflict. It is certainly the case that diaper-service companies have reported in recent years a significant upturn in business, no doubt reflecting concern in some quarters about the environmental impact of disposable diapers. And, as noted, both the diaper-service and the disposable-diaper industries have sponsored a series of studies to evaluate the various consequences of using one type of diaper or the other. These studies are all examples of what is known as "product-lifecycle analysis" or "cradle-to-grave analysis," a controversial and slippery methodology in which an attempt is made to gauge a full range of costs, in terms of energy use, pollution, public health, and money, that arise from the creation, use, and disposal of a product. This may include the cost of the gathering of the raw materials used in the product, of the transport and processing of those materials, of the act of manufacturing, of the packaging, of getting the product to market—the cost, that is, of virtually every conceivable element of every stage of a product's life cycle. The procedures and variables that ought to be involved in a product-lifecycle analysis are not yet standardized, however, and perhaps never will be. As you might imagine, the series of dueling studies that has been produced on diapers consists of fat tomes with bulky sections devoted to methodology, along with page after page of numbing tables and charts that can, after not very long, almost desensitize one to the contentiousness of the issues involved.

The first major salvos by both parties were fired in the late 1980s, after years of low-level skirmishing. On the anti-disposable diaper side, the key documents are *Diapers in the Waste Stream* (December, 1988), by Carl Lehrburger, and *Diapers: Environmental Impacts and Lifecycle Analysis* (January, 1991), by Lehrburger, Jocelyn Mullen, and C. V. Jones. Lehrburger, who is an environmental consultant, is the Torquemada of the anti-disposables coalition. His research has been underwritten in part by the National Association of Diaper Services, as already suggested, and by individual diaper-service companies, such as Baby Diapers, Inc., of Seattle, Washington, and Di-Dee Service, Inc., of Syracuse, New York. On the pro-disposable-diapers side, the fight has been waged primarily by Procter & Gamble and the American Paper Institute's Diaper Manufacturer's Group. Their proxies have been the accounting firm Arthur D. Little, which prepared the report *Disposable versus Reusable Diapers: Health, Environmental, and Economic Comparisons* (March, 1990), and the environmental-consulting firm Franklin Associates, which prepared two companion reports with similarly ponderous titles (both published in July, 1990). It is perhaps noteworthy —noteworthy, that is, either of a jockeying for position or an unwillingness to be put on the defensive—that all of the diaper studies, regardless of the authors' orientations, bear a notice indicating that the document has been printed on recycled paper.

In considering the plusses and minuses of disposable and cloth diapers, the major bones of contentions have involved these matters: the relative volume of discards; the relative amount of energy that the use of each diaper requires; the relative volume of raw materials that the use of each type of diaper requires; the relative amounts of water consumed; the relative threat of ground and water pollution; the relative threat of air pollution; the relative threats to public health; and the relative cost per diaper. Let's look at the contours of the situation that almost everyone can agree on.

For disposable diapers, the bulk of the energy use occurs during manufacturing, and at this stage there is also a likelihood that some pollution will occur. The resources required are mostly renewable—cellulose, from trees—but plastic, too, goes into the diaper, and it goes into the packaging as well. The manufacturing process requires large amounts of water, some of which becomes waste water. Disposable diapers obviously create more municipal solid waste than cloth ones do, and they create a possible pollution problem when they are dumped in landfills (a third of all diapers contain fecal matter, and all contain pathogens, at least initially) and perhaps even when they are incinerated (in the form of small amounts of chlorinated oxygen compounds). As for expense, if one simply looks at per-diaper cost, disposable diapers drain the pocketbook faster than cloth ones do (the per-diaper cost for disposables is about twenty-five cents, versus 7 to 9 cents for cloth diapers laundered at home and 13 to 17 cents for diapers from a diaper service).

For cloth diapers, the largest amounts of energy are consumed in the growing of the cotton (which requires large quantities of irrigated water and pesticides) and then in the 180 or so launderings that the average diaper laundered at home goes through in its lifetime. Diapers last more than twice as long at home as they do in the employ of diaper services (which only about 15 percent of households on a cloth-diaper regime use), largely because diaper services, for aesthetic reasons rather than purely practical ones, limit the number of times they will reuse a cloth diaper. The material resources required for participation in a cloth-diaper system (cotton, primarily) are almost completely renewable—but don't forget about the chemicals used to make detergent (or the ones used to grow cotton, for that matter). The washing of cloth diapers requires vast amounts of water and turns the water filthy; it all goes into the sewage system. Diaper services, because of economies of scale and other efficiencies, use less energy per diaper and produce less dirty water per diaper than is the case with home laundering, but the amount of both is nonetheless very considerable. As noted, the use of cloth diapers is relatively low if one looks simply at per-diaper cost.

So much for the areas of general agreement. The disagreements come when researchers take this general template of diaper reality and try to find, or generate, data to plug into it. The result is invariably a collision of assumptions and methodologies that yields widely disparate conclusions. A look at one major aspect of the two studies done by Franklin Associates in 1990 and the Lehrburger study of 1991—that is, the question of energy usage—shows how wide the disparities can be.

Considering the product's entire life cycle, does a disposable diaper result in the expending of more energy than a cloth one? Lehrburger finds that, on a per-diapering basis, disposables use six times more energy altogether than cloth ones do. Franklin finds that cloth diapers use twice as much as disposables. There are a number of reasons for the irreconcilable difference, some of them arcane. For example, because people who use cloth diapers frequently double them up, the number of disposable diapers used per child per changing (one) is less than the average number of cloth diapers used per child per changing (somewhere between one and two). But precisely what is that second number? Is it 1.2? 1.8? Ultimately, the answer will determine estimates of numbers of loads that must be washed, which helps determine energy needs. Franklin fixed the number high — at 1.79 per changing—which has the effect of adding to the energy costs of cloth diapers. Lehrburger fixed it somewhat lower—1.2 for diaper-service diapers, 1.8 for home-laundered diapers, for a weighted average of 1.72—which has the effect of adding to the comparative advantage of cloth diapers. (The evidence for any of these numbers is tenuous.) Lehrberger also didn't figure the energy used for transportation into his calculations—for example, the fuels consumed in transporting the cotton to manufacturers and then trucking the finished diapers to retail outlets, or the fuels consumed in shipping diapers to the United States from places like China (where a lot of cloth diapers are made). Lehrburger also didn't count the energy used to extract, transport, and process the fuels that would be consumed during transportation. Franklin, for its part, gave the disposable diaper industry a specious energy "credit," because some otherwise useless byproducts of the proccess of manufacturing the diapers' cellulose padding can be burned as fuel; this fact, of course, does nothing to diminish the amount of energy actually required to make the diapers.

In sum, it is hard to know just how to compare opposing studies on diapers. It is also hard to decide whether the information would really matter all that much if one could. Regardless of which type of diaper requires the most energy, the overall amount of energy under discussion is not very large—all told, in the case of disposable diapers, from raw material to final disposal, about 3.5 million Btu (British thermal units) per 1,000 diaperings. That amounts to 560 trillion Btu for all disposable-diaper diaperings in America a year—a figure that sounds high but in fact is relatively small given the minuscule size of a Btu (which is the amount of heat required to raise the temperature of a pound of water from 60 to 61 degrees Fahrenheit at sea level). In real terms, and using high-end estimates for both energy consumption and number of diapers worn, all the energy invested in the disposable diapers that a typical child uses in a year is equivalent to about fifty-three gallons of gasoline. That is the amount of gasoline that would be consumed by driving from Boston to Little Rock. We may never determine conclusively which kind of diaper, all things considered, is the more energy efficient, but neither kind is a major drain on our nation's energy resources.

The same quizzical quality that afflicts the findings about energy afflicts most of the other findings as well. One study claims that home-laundered cloth diapers are responsible for twice the atmospheric emissions of disposable diapers. Another study says that, no, the level of emissions is comparable. One study says that home-laundered cloth diapers consume four times as much water as disposable diapers. Another study says that, no, "single-use diapers use greater volumes of total water on a per-diaper-change basis." One study says that using cloth diapers is cheaper. Another study says, no, that's not the case—not if you calculate the time spent doing laundry and factor in the cost of that time if it were being remunerated at minimum wage. The studies agree that disposable diapers result in more solid waste than cloth diapers do, and that cloth diapers are responsible for more raw sewage than are disposable diapers. But the debate too often seems merely academic, with the contenders bogging down in minutiae like medieval scholastics, arguing over, say, how much more energy is created by burning, in an incinerator, a dry cloth diaper or a sodden Pamper, or over what percentage of people who use disposable diapers thoroughly rinse the soiled ones out in the toilet before throwing them away—a crucial component, as it happens, in the estimation of the variable amounts of water use under disposable-diaper and cloth-diaper regimes. (For the record, Franklin estimates that 5 percent of disposable-diaper households rinse the disposable diapers before disposing of them; Lehrburger uses an unlikely figure—50 percent—in his calculations. Although the fact is not widely known, nor the injunction heeded, the official position of the makers of disposable diapers, stated on the packaging, is that such diapers ought to be rinsed in the toilet before being discarded.)

Let us, however, be frank: Those who are concerned about disposable diapers are not concerned simply because they believe that disposable diapers are perhaps marginally more wasteful, in terms of energy waste or water consumption or air pollution, than cloth diapers (which they may or may not be). People are accustomed to accepting—and justifying—tradeoffs, particularly when they involve trading a certain souring of the environment for a palpable measure of convenience. Garnering the benefits of electricity has necessitated a compromise of this very kind. The advantages that disposable diapers offer—in time, in freedom, in aesthetics—are immediately apparent; disposable diapers are not an acquired taste. And in the eyes of their users, or when seen in comparison with the negative consequences of electric-power generation, the downside of disposable diapers seems to pale to insignificance.

Or, at least, it does unless the most prominent contentions of those who oppose disposable diapers have merit: namely, that disposable diapers are a prime contributor to the filling-up of the nation's landfills, and that their presence in landfills constitutes a potent new threat to the public's health—not only the health of people who work with garbage but, through groundwater, the health of whole communities. These concerns are the big two. And, in fact, there isn't much to them.

Let's start with the filling-up-of-the-landfills issue. Some startling numbers *do* get bandied around. The Portland *Oregonian* reported in 1987 that disposable diapers made up one-quarter of the contents of local landfills. One federal official, the director of the National Research Council's Commission on National Statistics, declared recently that disposable diapers "constitute 12 percent of total trash," a statistic that he attributed to the National Institute of Environmental Health Services; which the NIEHS attributed to the National Research Council's Board on Environmental Studies and Toxicology; and which the Board on Environmental Studies and Toxicology attributed to the Environmental Protection Agency's Office of Policy Planning and Evaluation (cautioning, by the way, that the figure 12 percent should be preceded by the words "up to"). We have already encountered the weirdly mercurial "16 to 32 percent" estimate, which was cited by Martha Gray at the Children's Museum, and appeared originally in the Puget Sound *Sound Consumer*. Studies by

Carl Lehrburger and other critics of disposable diapers, along with news reports from newspapers around the country, always hammer home the same point; to quote Lehrburger on disposable diapers: "No other single consumer product—with the exception of newspapers and beverage and food containers—contributes so much to our solid waste."

That statement seems like quite an indictment, though in fact it may be a little like saying that birds would be the biggest animals on earth if there were no mammals, reptiles, or fish. The Garbage Project has paid a lot of attention to disposable diapers over the years, and has consistently found that they constitute an average of no more than one percent by weight of the average landfill's total solid-waste contents deposited between 1980 and 1989 and an average of no more than 1.4 percent of the contents by volume. These findings have been confirmed in excavation after excavation. The range of diaper weights as a percentage of total garbage on all Garbage Project digs has varied only from 0.59 percent to 1.28 percent. The range of diaper volume has varied only from 0.53 percent to 1.82 percent.

Disposable diapers may be a big-ticket item in landfills compared with toothpicks and check stubs, but they are simply not in the same league with paper of various kinds (newspapers, especially, as Lehrburger notes), or items like ferrous metals and construction and demolition debris that are not "consumer products" but fill up landfills at a rate which is orders of magnitude greater than that of diapers. Given all the other, larger targets of opportunity, it may be misguided to draw a bead on disposable diapers. It is certainly an illusion to believe that eliminating disposable diapers would have anything but an imperceptible effect on the larger garbage picture.

As for the possibly deleterious effects of landfill diapers on public health, the issue does not appear to merit great concern. To begin with, even if disposable diapers do represent a problem, their addition to a landfill does not suddenly poison an otherwise pristine environment. The so-called "bioload" of a typical landfill—the census of its microorganisms, many of which are pathogenic—is so enormous that the contributions made by diapers seem relatively insignificant. Landfills already receive about 20 percent of the sludge from America's sewage treatment plants. They receive 8 percent of the septage from the country's septic tanks. Normal household garbage fairly brims with food waste, with the residues of personal hygiene, with pet feces. Medical waste of every imaginable kind finds its way into landfills, even though much of it should not. Even in a universe without disposable diapers, landfills would potentially serve up a puissant pathogenic bouillon.

Of course, our universe does have disposable diapers, and it is worthwhile figuring out whether adding them to landfills makes any noticeable difference. Trying to do so seems to have turned into one of environmental science's minor cottage industries. In the past fifteen years several scores of studies have been focused on the impact of disposable diapers and their contents on the landfill biosystem. The findings of the overwhelming majority of these studies, many of which involve the use of "lysimeters"—metal containers with a diameter of about six feet and that stand some seventeen feet high, which hold carefully controlled and monitored simulations of landfills—yield a picture of diapers in landfills in which most of the microorganisms the diapers contain die off over time, and the few microorganisms that do not die tend not to migrate very far.

The propensity of bacteria and viruses in diapers to expire in landfills has been widely documented. One scientist who has done so on the Garbage Project's behalf is the microbiologist Charles P. Gerba, who has analyzed some two hundred diapers excavated from various strata of several urban landfills (including the one at Fresh Kills, New York, and landfills in Tucson and in Naples, Florida). Gerba is a scientist whose theoretical concerns have never deflected him away from problems of decided practical consequence. Once, after a cross-country trip had planted the question in his mind, he conducted a study to see whether the cleanliness of a motel room—as determined, for the purposes of the study, by fastidious biological examination—bore a direct relationship to its price, and if so, whether the relationship was the one you would expect. (It did, and it was.) For his Garbage Project investigation Gerba subjected the excavated diapers to a set of extraction procedures designed to detect the presence of live pathogens. He was looking specifically for various enteroviruses, hepatitis A, rotavirus, Giardia, and cryptosporidium. In the end, Gerba found evidence of live virus on only a single diaper, and it was unclear whether the virus had come to the landfill inside the diaper or had contaminated the diaper after its arrival. When leachate from lysimeters that contain diapers is compared with leachate from "control" lysimeters that contain no diapers, the leachates are virtually indistinguishable.

Why do pathogens die off? The many studies that have been conducted thus far suggest that the answer involves several things. One is simply the acidity of the urine that most diapers contain, which kills some microorganisms. The heat at certain depths inside landfills can also be lethal to some pathogens. And, ironically, the toxicity of the leachate that landfills generate—and about which there is so much legitimate concern—does go to bat for society in at least one respect, making a landfill's interior environment most inhospitable. Taken together, the consequence of these and other factors is that the presence of live pathogens declines rapidly with the passage of time.

Given all the above, what should be America's stance on disposable diapers? The stance of many in public life, to judge from activity in state legislatures, is to Do Something. In 1990 some twenty-two states considered legislation involving disposable diapers. Some of this legislation had to do with prohibiting daycare centers from insisting that parents bring only disposable diapers when they drop off their children. That stipulation has long been the rule throughout the country—for obvious reasons from the point of view of child-care providers. But in the absence of any compelling public need, parents deserve freedom of choice.

Other legislation would provide tax incentives to use cloth rather than disposable diapers, either by levying a tax on the disposables or by exempting diaper-service companies from sales taxes. Whether a few pennies a day will persuade *Homo economicus* to alter established behavior is another question. States and communities certainly have the right, if it is the people's will, to levy taxes on whatever they please. It is to be hoped that they realize that doing so in this instance will not be solving any local garbage crises.

Several states, including Florida, Pennsylvania, and Vermont, have considered banning disposable diapers outright. A number of states have toyed with the idea of requiring an environmental warning on packages of disposable diapers. Legislation to this effect proposed in New York State would require the following label:

> Disposable diapers may take over one hundred years to degrade in a landfill. This product has significant environmental impacts and may pose problems in disposal. Disposable diapers are used once and discarded. This product will create significant disposal costs to your community if used regularly. You may wish to consider alternative products that have less impact on the environment.

The legislatures in nine states in 1990 considered banning any disposable diapers that were not biodegradable, though such legislation did not pass anywhere. It probably never will. Consumer goods made out of "biodegradable plastic" enjoyed a brief vogue until many of them came to be seen, correctly, as little more than products of a marketing scheme designed to tap into a perceived increase in "green" sentiment in the country. Although biodegradable plastic eventually can fall apart (but usually only with the help of sunlight, a scarce commodity in landfills), its constituents retain much of their bulk and hence take up as much room in landfills as regular plastic. Indeed, they may take up more room. The substance introduced into plastic that allows microorganisms to process it and therefore make the plastic degrade is typically cornstarch, but the cornstarch so weakens the plastic that extra plastic must be used in any given biodegradable plastic product to ensure that it has the same qualities of strength as those of whatever it is replacing.

There is no assurance, of course, that biodegradable plastics really will break down. As we have seen, biodegradation is not a landfill's forte. Whether the cornstarch in biodegradable plastics turns out to appeal to bacteria in landfills also remains to be seen. Archaeologists can certainly attest that bacteria have sometimes shown little taste for corn under far more appealing conditions. More to the point, the Garbage Project has unearthed from landfills dozens of cobs with kernels intact, some of them decades old. Be that as it may, even if the biodegradable plastic does break apart, the plastic doesn't disappear; it simply turns into many pieces whose aggregate volume is nearly the same as the volume those pieces had taken up as an assembled whole. Making products out of biodegradable plastic also undermines efforts to encourage the recycling of standard plastic, because the cornstarch is a contaminant.

The flaws in the idea of biodegradable plastic (as advanced thus far) are legion, and they were quickly seized upon by an unlikely alliance of environmentalists and the leading manufacturers of traditional disposable diapers, which denounced the new products as a delusion. A study on biodegradable plastics released by the organization Greenpeace quotes a spokesman for the Mobil Chemical Company making essentially the same point. "Degradability is just a marketing tool," the spokesman said in an interview with the Talahassee (Florida) *Democrat.* "We're talking out of both sides of our mouths because we want to sell bags. I don't think the average consumer even knows what degradability means. Customers don't care if it solves the solid waste problem. It makes them feel good." In 1991, in a move that will surely serve as precedent elsewhere, Minnesota led nine other states in ordering the manufacturer of Bunnies disposable diapers to remove the "Biodegradable" label from boxes of diapers sold in local stores.

In any event, leaving biodegradable plastic aside, it is to be hoped that whatever steps governments decide to take with respect to disposable diapers are not taken in the expectation that some local garbage crisis will be solved. It won't be. Indeed, the most palpable effect of restrictions on disposable diapers will be the effect on the way parents live their lives. And that should be cause for concern.

In thinking strategically about how to cope with garbage problems, policymakers need to be realistic. This in turn requires that each type of refuse be seen and evaluated in behavioral perspective. In the case of disposable diapers, people are not prepared to be pushovers. It is said that lions, once having tasted human flesh, willingly eat no other. The parents of infant human beings display similarly strong preferences when it comes to disposable diapers. *Everyone* believes or suspects that disposable diapers are not the greatest thing for the environment, but the great majority of all parents today use disposable diapers either exclusively or in select situations, and are prepared to live with whatever guilt they may feel.

Disposable diapers are only one of many modern products that make life easier for countless millions of people. As the historian Samuel Hayes has pointed out, it was the advent of an easier life, with more free time for America's vast middle class to enjoy their surroundings, that gave rise to an environmental movement in the first place. Convenience and leisure on the one hand and concern about the environment (and garbage) on the other are inextricably linked. An erosion of the former will result in an erosion of the latter.

The disposable-diaper issue has surely prompted many people who might never have given much thought to garbage matters at last to focus on them. Its symbolic power has properly helped to raise the general level of concern. But as we attempt to cope rationally with America's garbage, we should avoid mistaking the most convenient symbols for the most pressing problems. And we should pick our battles carefully.

PART IV

Garbage and the Future

CHAPTER 8

THE TECHNOLOGICAL FIX

We have emphasized now and again how garbage studies can provide an important corrective to other ways of investigating the personal behavior of people in groups. However, garbology has its own built-in problems, chief among them the inconvenient fact that the garbage coming out of a household rarely represents all the garbage that that household produces. People pack lunches and go off to work or school, throwing the garbage away there. A significant if unquantifiable amount of food waste is diverted away from the solid-waste stream and into the mouths of dogs and cats. There is no doubt an infinite variety of reasons to explain why some portion of what *could* have been in a household's garbage doesn't make it into the bag or can, and the result is that many Garbage Project estimates for such things as patterns of consumption of certain kinds of foods tend to be on the conservative side. It could be that, had all the raw (and cooked) data been available, certain trends might have appeared more pronounced than they actually did.

As noted earlier, one of the main contributors to this form of bias is the kitchen garbage disposer, a technology that, while available commercially since the 1930s, did not become a standard feature in new homes until the 1970s. The first garbage disposer designed for use in a household kitchen sink was a descendant of the large grinders and shredders that municipalities employed beginning in the 1920s to prepare some solid waste for disposal in municipal sewer systems. The household garbage disposer came on the market in 1935; it was twenty inches in length, weighed seventy-five pounds, and bore General Electric's trademark. Although the Second World War delayed the device's refinement, in the postwar years other companies joined General Electric in the garbage disposer business, and the machines themselves grew smaller and lighter—and more appropriate for widespread household installation. Enthusiasts like Morris M. Cohn, a conscientious public servant in Schenectady, New York, and the editor for many years of the garbage-industry journal *Wastes Engineering,* claimed that garbage disposers would eliminate garbage cans the way flush toilets had eliminated outhouses. Cohn, whose books include *Sewers for a Growing America* and *By the Magic of Chemistry: Pipe Lines for Progress,* had begun tirelessly promoting the idea of a household garbage disposer in the early 1930s, and it was largely as a result of his encouragement that General Electric took the steps that led to its introducing the first commercial model. Although not directly involved in the engineering of the device, Cohn certainly deserves the title "Father of the Kitchen Garbage Disposer."

Cohn's remarks in an article in *Sewage Works Engineering* make plain that he heartily approved of the actions of the town of Jasper, Indiana, which became the first community in the United States to vote to place itself entirely in the hands of this new technology. As the historian Suellen Hoy recounts in a 1985 article titled "The Garbage Disposer, the Public Health, and the Good Life," which was published in the journal *Technology and Culture* in August of 1950, this town of 6,800, with a bothersome open dump and a recent history of hog cholera that had been traced to infected slops, set about installing General Electric garbage disposers in all of the town's household kitchen sinks; at the same time, Jasper discontinued all public collection of wet garbage and prohibited the discarding of wet garbage in garbage cans. "Somebody has to stick his neck out and do things like this," said Jasper's mayor, Herb Thyen. "Otherwise progress ceases." By October, the new technology was in place everywhere, and Jasper began its new life as a town without a garbage collector. The initial results were encouraging. There was no deleterious effect on the sewer system, as some had feared, and there were fewer flies in town (according to a before-and-after "flies per grill" count made on automobiles). As a side benefit, Suellen Hoy reports, the installers of garbage disposers found and corrected numerous cases of defective amateur wiring. General Electric began distributing a brochure whose cover featured a young boy looking up at his father (book open on knee, pipe in hand) and asking: "Dad —what was garbage?" Of course, the collection of non-wet garbage would still be necessary, but the universal availability of disposers to deal with organic household debris would keep the volume to a minimum.

Inspired by Jasper's example, a number of other communities in the Midwest took up what became known as "the Jasper plan." And, it must be said, the efficient disposal of garbage was not the sole impetus. Garbage disposers promised not only to get rid of garbage, more or less effortlessly, but also palpably to improve the quality of life. The garbage disposer symbolized the American ideal. "In essence," Suellen Hoy writes,

> this "hunk of better living" touched a responsive chord in a generation of Americans who, having survived years of Depression grayness and wartime scarcity, resumed their search for a healthier environment and a "greater ease of living" through goods and amenities that offered more cleanliness, convenience, and comfort.

The disposer has certainly made life easier, but it turns out not to have made all that much difference as far as garbage-generation rates are concerned. We checked up on Jasper, Indiana, not long ago, and spoke with Jasper's street commissioner, Robert Main. How was the future going? Well, he said, Jasper still didn't have anyone picking up wet garbage, and it still gave out tickets to people whose trash cans were found to harbor such garbage. But the town had never been able to dispense with a pickup of non-wet garbage. Now Jasper's landfill was nearly filled, Main said, and the town had had to ask the state to allow it to pile refuse higher and higher. Insofar as garbage is concerned, Jasper is now scarcely distinguishable from anyplace else in the United States.

By and large Americans have never been content to do things the old-fashioned way, and where garbage has been concerned they have always been receptive to any new state-of-the-art means of disposal —to each new technological fix—especially if it promised a savings in money (Fire the garbage collector!) or, better yet, a tidy profit. In the mythology of the American Dream, the relationship between advancing technology and a state of personal well-being that ratchets ever upwards was long assumed to be linear and direct. And, until recently, this assumption seems to have been stunningly unaffected by the repeated failure of technological fixes to perform precisely as advertised. Today, of course, technological backfires and misfires, real and alleged, have become so common that the old mythology is at best unchic, at worst an object of hostility. That the pendulum has swung in this direction is perhaps not a bad thing. One lesson of the Jasper story may be that ambivalence is the most sensible stance to take toward many technological innovations, including those that involve garbage. Such a stance may allow us to employ realistically the technological tools that we possess or may develop.

The history of the technology of garbage disposal essentially begins with two industrial techniques—incineration and reduction—that have their roots in the nineteenth century and that were exported to America from Europe. Of the two, incineration was the earliest to appear and, despite initial setbacks, has proved the more durable.

The first garbage incinerator, known as a "destructor," went into operation in Nottingham, England, in 1874, and the technology took scarcely a decade to cross the Atlantic. The U.S. Army built the first American model, called a cremator, on Governor's Island in New York City in 1885, and during the next few years cremators were fired up in Wheeling, West Virginia; Allegheny, Pennsylvania; and Des Moines, Iowa. As the historian Martin Melosi has noted, cremators enjoyed an initial vogue until their expense (wet garbage had to be combined with coal, or it wouldn't burn) and the incomplete combustion of garbage that they achieved began to cause widespread dissatisfaction. "Of the 180 furnaces erected between 1885 and 1908," Melosi writes, "102 had been abandoned or dismantled by 1909." By 1920 cremators were still operating in only about a dozen American cities.

The Age of Incineration would one day return, but it was first necessary for an Age of Reduction to enjoy a brief vogue. Reduction was a technique that evolved in part out of the whaling industry (where it was used in the rendering of blubber) and involved stewing wet garbage and dead animals—there were fifteen thousand dead horses a year to get rid of in New York City at the turn of the century—in large vats in order to produce grease and a substance called "residuum." The grease was sold for between three and ten cents a pound and was used in the manufacture of soap, candles, glycerine, lubricants, and perfume. The residuum brought between five and ten dollars a ton, and was used for fertilizer. The waste-management historians Rudolph Hering and Samuel Greeley date the first reduction plant in the United States to the year 1886; their modern counterpart, the historian Martin Melosi, dates it to 1896.

Regardless, it was New York's Colonel Waring who gave the reduction process broad public impetus. Waring, a man who would, one suspects, feel quite at home amid our present-day garbage wars, at one point tried to get the people of New York City to participate in what today would be called a curbside separation program—leaving out trash with all the glass, paper, and "wet" garbage in separate containers—so as to be able to recycle the glass and paper, but he soon gave up the idea when New Yorkers proved, in his words, "obdurate." In 1896, determined to give the recycling of garbage another try, Waring entered into a contract with the Sanitation Utilization Company, an enterprise specializing in the process of reduction. The Sanitation Utilization Company turned a handsome profit, and, owing to Waring's redoutable national influence—he was something of a cross between H. Norman Schwarzkopf and C. Everett Koop—reduction plants were soon in operation throughout the country.

One reason for the success of reduction in the United States was that, by world standards, this country was such a rich one; its garbage, therefore, was also rich, and so, in the end, were the liquids that could be distilled from it. The negative side of reduction was that reduction plants emitted nauseating odors as well as a black liquid runoff that polluted nearby watercourses. A reduction plant was a truly foul industrial enterprise, of a kind that most Americans can no longer either remember or imagine, much less tolerate. After several decades in the ascendant, reduction plants began to close. Most had been shut down by the beginning of the Great Depression. The very last one, in Philadelphia, ceased operations in 1959.

As reduction waned, incineration made a considerable comeback, and by the eve of the Second World War some seven hundred incinerators of improved design were in operation in the United States. The disadvantages of incinerators were well known from the start—they discharged foul odors, noxious gases, and gritty smoke. But the chief advantage of incineration was overwhelmingly compelling: It burned garbage to a crisp. Or, at least, almost to one. Depending on the method, incineration leaves a residue of ash that may amount in volume to between 5 and 15 percent of the volume of the garbage that has been burned, and the ash must be disposed of—often in municipal landfills.

In 1988, during excavations at the Sunnyvale landfill, near San Francisco, the Garbage Project learned that for many years prior to becoming a landfill, beginning perhaps in the 1920s, Sunnyvale had been the site of an open dump. A city engineer, who as a youth had taken target practice there, produced a map he had made of the area in 1957 and indicated the place where, if the Project dug, the remnants of the original dump would be found beneath the modern landfill. The prospect of turning up garbage from the 1920s through the 1950s was tantalizing, and the bucket auger was positioned accordingly. But when it reached the appropriate depth, what ended up being retrieved was bucket after bucket of incinerator ash.

Like reduction, incineration enjoyed a popularity, if that is the word, of several decades' duration, only to fade during the postwar years in the face of competition from the sanitary landfill, which to many minds seemed to represent the best technological fix yet. Decades before the public had heard of "risk factors" or acid rain, municipalities in the 1950s began shutting down their incinerators, with the blessing of a local citizenry whose respiratory systems and aesthetic sensibilities the incinerators at times offended. Within a few years of the passage of the Air Quality Act of 1967 and the Clean

Air Act of 1970 there were, by one account, only about 150 garbage-burning incinerators left in the United States.

Then came the energy crisis, and soon afterwards an object lesson in why making policy on the run during moments of upheaval so often leads to undesirable results. Amid skyrocketing fuel costs and a fear of resource shortages, incinerators were reconceived, retooled, and renamed: They became "resource-recovery facilities," and building one came to be seen almost as a patriotic duty. (The change in terminology was no accident; perhaps, a *Waste Age* editorialist wrote in 1970, "the time has come to drop 'incinerator' and the picture of the past which it frequently calls to mind.") The idea behind resource-recovery plants was simple: Not only would they burn garbage up, they would also provide heat or electricity to customers nearby. Additionally, the resource-recovery facilities would be equipped with sophisticated pollution-control devices. As with many other forms of garbage technology, the resource-recovery concept was essentially imported from Europe, where demographic congestion in many places does not permit a reliance on landfills as the primary means of garbage disposal.

Resource-recovery facilities arrived in two basic forms. The form that was initially the most popular is what is called a refuse-derived fuel facility (RDF). Solid waste is fed onto a conveyer belt from a storage area, pulverized by swinging hammers in a shredder, stripped of iron-based metals by magnets and of aluminum by blowers, and cleaned by being passed through screens that allow all the sand, glass, rock, and other noncombustibles to drop out. Everything that makes it through this gauntlet is then shredded again to become the eponymous refuse-derived fuel, which is sold to power companies as a papery fluff or in the form of pellets (called "densified" RDF), or is burned on-site to create energy. A 1977 article in *Reader's Digest* had this to say about RDF plants: "Though the pioneers ran into many difficulties—par for any technological change—the problems of recovering energy and raw materials from garbage now seem to be solvable."

Actually, with respect to the first RDF facilities, they weren't really solvable. The fact is, RDF plants tried to accomplish too much. A surfeit of variables had to be factored into their operations. For one thing, the machinery was complicated and delicate. The hammer mills in the shredding compartment repeatedly fell victim to such mundane antagonists as pantyhose, which fouled the works, and pressurized containers and flammable liquids, which contributed to explosions. The plants were expensive to operate and partly dependent for their profitability on sales of the recyclable materials that they recovered, but the secondary-materials markets are exceedingly erratic and, in any event, already serviced by a small army of professional traders. It is at once ironic and typical that the March 15, 1974, cover story of *Science* magazine, which evaluated the potential for refuse-derived fuel, and included in its evaluation a discussion of costs and revenues, did not address the subject of fluctuations in the commodities markets even in passing.

The plants were dependent as well on sales of refuse-derived fuel, and many of their customers, such as the operators of coal-fired power plants, complained that it often contained too much moisture. In a lot of cities, officials eventually realized, the weather occasionally brought rain or snow, or both, and not everyone was careful about keeping the tops on their garbage cans. The owners of many incinerators also found themselves locked in battle with local recyclers over garbage they both wanted, such as newspapers and cardboard, which burn easily and hot. Others had difficulty just getting enough garbage to operate—either their tipping fees turned out to be higher than those charged by landfills, or they had miscalculated the amount of garbage that would actually be generated locally. Some operators, like the Canadian company that ran the resource-recovery plant in Akron, Ohio, ended up having to import garbage from out of state. A commercial load of imported sawdust, oil, and paint wastes from Kearney, New Jersey, was fed into the Akron plant in December of 1984, resulting in a fireball that exploded out of the hammer-mill and killed three people.

Ultimately, more than a dozen of these exotic and expensive early RDF plants simply went out of business. But the technology has been reborn in smarter, sturdier, more streamlined fashion, and there are today about a dozen facilities in the United States that create RDF, and almost twenty that both create and burn it. There are a number of reasons to believe that more RDF plants will appear in the future. The new RDF facilities are more selective about the garbage they are willing to receive, which means that they keep such things as car batteries, electronic circuitry, and items containing mercury—all of which contribute heavily to toxic emissions—out of the furnace. They can also, but need not, separate out and recycle any commodity for which there is sufficient demand. If they produce more refuse-derived fuel than they can use themselves, they can sell it to power plants. And, of course, they generate electricity, which they sell to local manufacturers or to a utility. (The 1978 Public Utility Regulatory Policies Act, or PURPA, mandates that utilities must buy electricity from companies that produce it by means of waste incineration.) In terms of function, then, communities or consortia of communities that build RDF plants can essentially custom design the facility and its scope of operations.

The other major form of resource-recovery facility is called a "mass-burn" incinerator. It is a far simpler proposition than an RDF facility and far more widespread. The operators of a mass-burn incinerator need not (and usually don't) separate out materials for recycling in advance of incineration. The garbage is fed into a furnace where it falls on moving grates which tumble the garbage around at temperatures of 1800 to 2000 degrees Fahrenheit. The burning mass heats water in a centralized boiler or in tubes in the furnace walls. The steam drives a turbine to generate electricity that is sold to a utility. There are about fifty mass-burn incinerators in operation today in the United States, and fifteen more are under construction. Some forty mass-burn facilities are in the planning stage.

Incineration's modest recovery during the late 1970s and 1980s—its third coming, one might say—has occurred in the context of a fluctuating economic and political situation, and has met with the sustained attention of people concerned about the larger environment. The terminology has changed again: Incinerators are now "waste-to-energy facilities." The reception has not always been a friendly one. The Garbage Project's own experience with incineration has involved studies, commissioned by environmental consultants, of the hazardous constituents of the household waste that might find their way into incinerators, the idea being that exerting some control over what gets onto the conveyer belt may have some benign influence on what comes out the smokestack. The main concern these days about incineration of whatever kind—besides the matter of cost—obviously has everything to do with pollution.

Modern incinerators do have sophisticated mechanisms to curb pollution. Waste gases are blown from the burn chamber into acid scrubbers and then either through electrostatic precipitators or through fabric filters in what is known as the "baghouse." For all the precautions, well-run incinerators can release into the atmosphere small amounts of more than twenty-five metals and acid gases, as well as a class of chemicals known as dioxins, of which there are some seventy-five different kinds, and which have been implicated in birth defects and several kinds of cancer. Just how toxic some of these materials are in small doses is an issue embroiled in bitter controversy. For dioxins, the estimates of the human toll range from one death per every million people over a period of seventy years (the figure put forward by the Centers for Disease Control, based on "hypothetical, yet currently accepted, models of risk estimation") or two or three deaths per million people over the same period (the figure put forward by scientists working for the City of New York), to possibly ten or twenty or even more deaths per million people over a seventy-year period (the figures put forward by the environmental activist Barry Commoner). The size of the threat to public health, whatever it is, probably varies from plant to plant.

A portion of the ash produced by incinerators—the flyash, which follows waste gases through the exhaust and scrubbing system, as opposed to the bottom ash, which is heavier and falls through the grates—is toxic, usually containing dangerous levels of lead, cadmium, and dioxin, and some cities have had difficulty disposing of it. In September of 1986 the city of Philadelphia, which operates two mass-burn incinerators but has no landfills, loaded up a cargo ship, the *Khian Sea,* with sixteen thousand tons of incinerator ash and sent it forth in search of a dumping ground: an exercise in "waste imperialism," as one editorial writer called it. The ship was barred from every port it tried to enter over a period of two years. (It eventually turned up in Singapore without its cargo, the fate of which was not disclosed.) Most of the toxic incinerator ash winds up in ordinary landfills, thanks to an exemption it enjoys from the

Environmental Protection Agency. Whether that exemption will survive the coming reauthorization of the federal Resource Conservation and Recovery Act remains an open question.

Quite apart from health issues, incinerators are hugely expensive—it may cost $250 million to build a mass-burn incinerator capable of handling two thousand tons of garbage a day, which is roughly the amount produced by New York's Westchester County—and the task of getting one sited attracts the kind of pork-barrel chicanery and eco-bravado that one might expect. The protagonists usually pit best-case scenarios (the plant will work as designed and maintain high, even temperatures that will destroy hazards before emission) against worst-case scenarios (faulty design and operator error will result in large and deadly discharges of toxic substances). There is, of course, something to be said for both sides. Plants can usually be operated safely—for a time. But plants get old, and performance will begin to decline unless the plants are modernized. Nevertheless, incineration is a significant piece of the future—it handles some 15 percent of all municipal solid waste already, and may handle 25 percent by the year 2000—and even some environmentalists, or at least some mainstream ones, seem more or less resigned to incineration as an unavoidable necessity: the only conceivable alternative to landfills for at least some portion of our garbage.

In the end, the question of siting and building incinerators will be a political one. Communities must decide whether the possible risks, such as those embodied in the estimates cited by Barry Commoner or those embodied in the lower estimates cited by regulatory agencies and the builders and operators of incinerators, seem acceptable. American society now and in the past has certainly proved capable of embracing technologies far more threatening than those embodied in a modern incinerator. The advantages brought to us by the automobile, we now know, come at a cost of 46,000 lost lives every year, and many times that number of devastating injuries; the price we pay for the automobile in terms of pollution has, of course, been staggering. Had we known in advance that this might be the case, would production of this machine ever have been allowed to proceed? Should it have been?

We have managed to make peace with countless other less devastating but still harm-doing manifestations of progress, accepting them because the good that they accomplish is seen to outweigh by degrees of magnitude the problems they cause. One thinks, for example, of X rays, say, or vaccination. There is a difference, of course, between risks incurred voluntarily and risks incurred involuntarily, though the line between the two is in places hazy, and drawing it does not affect relative hierarchies of risk, so far as they can be known. Is incineration better than a big new landfill? Certainly *something* must be done to accommodate what recycling and source reduction can't cope with (whatever percentage of the solid-waste stream that turns out to be), and communities will make decisions taking into account the particulars of their specific circumstances. The decisions will vary accordingly, as they should.

The circumstances facing some thirty-two communities in southeastern Massachusetts and on Cape Cod during the 1980s led them ineluctably, though with detours into legal, financial, and political wrangling, to a decision to build a waste-to-energy facility that would serve them all (and by now, nine other communities). The SEMASS facility, located in Rochester, Massachusetts, has been in operation since 1988, and it processes the garbage generated by more than a million people—nineteen hundred tons of garbage a day in all. The forty-one communities now served by SEMASS occupy a tract of geography where the water table is extremely high—much of the catchment area consists of the cranberry bogs that make the region the largest cranberry-growing area in the nation—and where there is much protected land. The region is scenic and a major vacation spot. For all that, the area is also relatively densely populated.

When the communities in southeastern Massachusetts began running out of existing landfill space, few were able to find the land, much less the political will or the hard cash, to create more. By offering the towns a guaranteed low tipping fee of about $12 per ton (it's about $20 per ton for new participants now), the owners of SEMASS, a consortium of five private companies—were able to entice them into long-term contracts. Local tipping fees at landfills at the time when the first contracts were signed, in the early 1980s, were about twice as high as what SEMASS was offering to charge.

Today, a portion of the garbage from Cape Cod arrives at SEMASS by rail, saving energy and cutting down local truck traffic; the railroad cars, each of which holds about forty tons of garbage, roll into the facility's cavernous, 170,000-square-foot "accepting bin" where they are picked up by a powerful device and turned upside down, the garbage spilling onto the floor. When shredded and burned the garbage creates steam that powers generators which provide some fifty megawatts continuously to Commonwealth Electric, satisfying about 6 percent of the utility's total needs. The SEMASS facility is cleaner, in terms of emissions, than is a typical Boston Edison or Commonwealth Electric power plant; regardless, then, of what one makes of its health risks in some absolute sense, by saving the public utility the need to generate a certain amount of electricity through conventional means, SEMASS helps make the air cleaner than it otherwise would have been. Because the possibility of groundwater pollution is such a large local concern, SEMASS is a "zero-discharge" facility, meaning that all the water it needs for its various processing and generating operations flows within closed systems. As one would expect, the SEMASS facility is closely watched by environmental groups, which are not shy about registering complaints.

In this age of Epcot, the idea of a factory tour no longer has the cachet or kindles the excitement that it once did in the United States, but many industrial establishments are willing to provide such tours, if asked. A modern incinerator is worth a look. At SEMASS the trains and trucks bring garbage to the floor of the accepting bin where personnel in orange jumpsuits check each delivery for unacceptable items (concrete slabs, spring mattresses, and overstuffed chairs, for example; much of the rejected material must be taken away by the hauler) before bulldozers urge the mountains of waste onto conveyer belts. High above, a picture window gives onto a control room, where an engineer watches the next legs of the journey on TV monitors. The garbage passes by a picking station where more items—tires, carpets, cables, car batteries, and such—are removed (they'll go to a landfill at the incinerator site), and then continues onwards into the shredders. After being shredded the garbage passes through magnets, which separate out any ferrous metal; this is carried off elsewhere by conveyer belt, eventually dropping through a funnel into the back of a dump truck, sounding like the cascade of ice cubes from an ice machine into a glass. The shredded garbage is then burned.

As noted earlier, two kinds of ash result. One kind, flyash, which is deemed hazardous, leaves the boiler with the exhaust gases, passes through a spray dryer and electrostatic precipitator, and is chemically stabilized to help prevent heavy metals from leaching out. Finally it is trucked off, wet, to a double-lined landfill, where it hardens to a rock-like consistency. The other kind of ash, called bottom ash, is heavier and bulkier, and consists of the burned residues of incinerated materials that fall through the grates in the boiler's floor. This consists primarily of a pumice-like substance known as boiler aggregate, or clinker, which at SEMASS is formed into high-quality cinder block. Portions of the facility's administration building are built out of this material. Much of the rest of the bottom ash consists of nonferrous metals of various kinds, including aluminum, which is sold to scrap dealers. In each ton of clinker and nonferrous metals that SEMASS collects from the bottom ash—and it collects about eight tons of it a day—there is more than $1,000 in coins, which would be redeemed at full value by the United States Treasury were it worth anyone's time to go through the vast metal bins looking for them. (One can only wonder about the formation processes that lead the million or so people served by SEMASS to leave $8,000 worth of coins in their garbage every day.)

As for emissions from the SEMASS facility's stack, they have so far been well below the strict limits set by the state's Department of Environmental Protection. The companies that operate SEMASS hope gradually to introduce some improvements in the whole process. One is to add a stage at the very beginning to allow for the separation of plastics, so that they could be recycled. Another is to find a safe end use for the flyash that currently must be landfilled. But even as currently run, the facility demonstrates that incineration need not be a reckless, thoughtless endeavor.

Incineration does not make sense everywhere, and there are those who will argue that it does not make sense anywhere. Given the potential dangers of the technology involved, on the one hand, and the size of the capital investment involved, on the other, it is perhaps not surprising that some people take doctrinaire positions on incineration, and then dig in their heels. The more extreme among the opponents of incineration will at times resort to the old standbys of protest movements in general, ranging from legal harassment by means of lawsuits to the physical obstruction of roads and entranceways. Industry spokesmen, for their part, have at times been too blithely dismissive of legitimate environmental concerns—too quickly discounting the levels of risk posed by incinerator emissions, offering glib avowals that eating a peanut-butter sandwich every day, for example, may actually be more harmful (owing to the presence of aflatoxin, a naturally occurring carcinogen) than living next door to a waste-to-energy facility.

Allen Hershkowitz is a senior scientist with the Natural Resources Defense Council. He has probably inspected more incinerators in the United States, Europe, and Japan than anyone else alive. He is an ardent proponent of recycling. His environmentalist credentials are highly burnished. Hershkowitz believes that incineration, if it is pursued with all the appropriate safeguards in place, is a viable component in some communities of an effective solid-waste-management strategy. Consider, he points out, the situation that New York City finds itself in today. In the aftermath of the passage of the Clean Air Act of 1970 some thirteen of the city's incinerators could no longer meet federal emissions guidelines. Three of the thirteen were able to be "retrofitted" with pollution-control devices, but the other ten had to be shut down. A plan was eventually developed to build five new incinerators, with the first of the five to commence operation in 1984. The plans, of course, were beset from the outset by community resistance, political bickering, and changing environmental regulations, with the result that none of the incinerators has yet been completed. Meanwhile, the garbage that had once been incinerated has been diverted to Fresh Kills landfill, which during the past two decades has released several billion gallons of leachate into New York harbor. Perhaps, in time, source reduction and recycling will be able to substantially diminish the volume of all that diverted garbage. But as pragmatists like Hershkowitz realize, source reduction and recycling are not a complete solution for the problems that are upon us now.

If the future is to hold a place for incineration, it ought also to require scrupulous adherence to the highest standards of pollution control, pollution monitoring, and ash disposal that technology can offer. But technology is not enough. One basic improvement would be simply making sure that those who have hands-on responsibility for running waste-to-energy and other such facilities have been professionally trained for the work they will be doing. At the moment there exists no set of national criteria for what those who run incinerators ought to know; the training programs in this country do not begin to compare with those available in Europe and Japan. This lack of training is, in a way, symptomatic of the broader American unwillingness to see the disposal of garbage (as opposed to the generation of electricity) as an important enough enterprise in itself.

"Some U.S. incinerator operators," Allen Hershkowitz has written, "try to hire workers with 'steam experience'—those who have operated coal- or oil-fueled power plants." But, as he goes on to point out, operating a waste-to-energy facility is not like operating other kinds of power plants. The fuel, for one thing, is heterogeneous, and sometimes wet; these qualities can give rise to the fluctuating temperatures in the burn chamber which, experts are beginning to agree, are in large measure responsible for the creation of dioxins. The point is, those who are responsible for running incinerators need to understand that the safe destruction of garbage is their primary business, with electricity-generation a happy secondary consequence. And they need to be given the tools to do the job.

The situation facing incinerators is in some respects uncannily similar to that faced by nuclear energy plants. Writing in *The Wilson Quarterly* in the summer of 1979 the physicist Alvin Weinberg criticized, among other things, "the mistaken belief among utility executives that a nuclear plant was just another generating station." And he went on:

> But the responsibility borne by the nuclear operator is so great that he and his staff must be regarded—and trained—as an elite. They must constitute a cadre with tradition, competence, and confidence.

Needless to say, a waste-to-energy facility will never pose the potential threat of a Chernobyl. But the esprit that Weinberg would have liked to see has much to recommend it.

The lesson that should be drawn from episodes like the one in

Jasper and the rest of the long history of garbage technology is that, much as we'd like to have one, there is, in fact, no foolproof technological fix. There are a number of helpful technologies, and we should use them, but they all have their limitations, and the claims for them should generally be kept modest. Nor should we expect them to turn out to be a license to print money or a means of accomplishing important social ends other than the primary one. Getting rid of garbage is one of those basic things, like maintaining public health, that we must learn to accept as being somewhat expensive, and must also learn to accept as a noble enough goal in its own right—worth doing even if one hasn't made a buck or saved the world in the process.

CHAPTER 9

CLOSING THE LOOP

In a typical household in the United States one can expect to find at least one example of each of the following: a washer and a dryer, a refrigerator, a stove, a stereo set, a television set, a couch, an armchair, a kitchen or dining-room table, one or more chairs to go with that table, a dresser, a bookcase, and a bed. Consider this minimal list of furnishings on a national scale and one confronts a stockpile of potential future discards of extraordinary size. The life expectancy of each of these items is, after all, finite. Those who conduct the kind of materials-flows analyses mentioned in chapter two generally assume, for example, that the service life of refrigerators and other major appliances is at most about twenty years. Thus, if it is known that in a certain year in the mid-1950s the United States produced 3.7 million refrigerators, 3.6 million kitchen ranges, 4.5 million washing machines, and 1.5 million dishwashers—which, in fact, it did, in 1956—then one can be fairly certain that by 1980, say, roughly the same number of appliances would be found entering the solid-waste stream. (A similar assumption can be made in the case of furniture.) Applying such assumptions and such numbers to a city the size of Tucson, one would have expected by 1980 or thereabouts to be faced with the need to dispose of some 24,000 discarded major appliances a year, or sixty-five a day.

The materials-flows assumptions seem to make sense. The only problem with them is that when the Garbage Project set out looking for major appliances in Tucson's solid waste, it couldn't find very many of them. This was not for lack of trying. Two Garbage Project researchers, Paul Freidel and Bruce Douglas, were detailed to Tucson's Los Reales landfill for a week and told to keep a record of all the major appliances and big pieces of furniture that were hauled to the site. As it turned out, few of the targeted items ever appeared. And, tellingly, every one of the few appliances and sticks of furniture that did arrive was carted off within hours by some other visitor to the landfill: One man's trash, apparently, is another man's treasure.

The extent to which big-ticket discards seem to get a new lease on life was even more powerfully underscored during an early-morning Garbage Project stakeout in advance of one of the special pickup days in which sanitation crews collect the bulky durable goods that they ordinarily shun. Before the garbage collectors arrived, scavengers in pickup trucks swept rapidly through the alleyways of Tucson. They methodically scavenged every major household appliance and piece of furniture in sight and ferried them off to larger trucks stationed nearby, which in turn carried the booty off to scrap dealers and resale shops and even to Mexico.

Michael Schiffer, an archaeologist at the University of Arizona, and a longstanding Garbage Project ally, decided to head up a study group that would look into the matter more closely. Schiffer had been interested for years in mechanisms of reuse—these provide an important component of archaeological "formation processes," which is one of Schiffer's specialties—and was a frequent visitor to thrift shops, yard sales, and swap meets. He and a handful of anthropology students, under the aegis of what they called the Reuse Project, began conducting interviews with members of randomly selected households in a broad transect that cut across all of Tucson, asking householders whether they had recently disposed of any one of the thirteen major appliances or pieces of furniture mentioned at the beginning of this chapter. In the 184 households surveyed, the interviewers learned, 743 of the specified items had recently been replaced. And what was the fate of those replaced appliances and pieces of furniture? A little more than 30 percent of them were still kept around the house somewhere; 34 percent had been sold or given to strangers or stores; and 29 percent had been sold, given, or loaned to relatives and friends. Only 46 of the 743 items (or 6.2 percent) had been thrown away.

Schiffer then turned the issue around, looking at it from another angle. He conducted inventories of the respondents' homes, making note of all the appliances and pieces of furniture that were said to have been acquired second hand. Where had *these* items come from?, he asked. The acquisition mechanisms broke down like this:

Gift from a relative	19.9%
Rented with a dwelling	19.5
Purchased at a new-used specialty store	11.9
Inheritance	9.5
Gift from a friend	5.8
Purchased with dwelling	5.6
Came with dwelling	3.3
Purchased from a friend	3.3
All others	18.0

The astonishing thing is that the total number of used durable items reflected in the data above was 2,412, for an average of more than 12.5 per household; used appliances and furniture represented fully one-third of all the major appliances and pieces of furniture in the 184 households surveyed.

For all the talk of the United States as a "wastemaker" society, the informal and commercial trade in used goods—most of which escapes the notice of the government's record-keepers, tax-collectors, and other official statisticians, and goes largely unremarked by the public as well—is apparently huge. Judging from Michael Schiffer's study, at any given time only a fraction of the appliances and pieces of furniture that households no longer need or want is being "thrown away" in any conventional sense of the term. The rest of these durable goods continue to live healthy and productive lives, laterally cycled for years from place to place, and eventually cannibalized or sold as scrap. Many end up south of the border. (One Garbage Project associate, Ramón Gomez, encountered operational Westinghouse washing machines from the 1940s, with attached roller-wringers, in households in Nogales, Mexico; it is a sobering comment on U.S. industry that in Mexico, among used appliances that are between five and twenty-five years old, the prices for many older American durables tend to be higher than for many newer ones.) Even the durables that do finally get thrown away—left out at night for disposal—are likely to be subject to the attention of alley scavengers and scavengers at landfills.

"Thus," Schiffer wrote in a summary of his findings, "it appears that few pieces of furniture and appliances reach the Tucson archaeological record [that is, the local landfill] intact." And he went on to observe, pointedly, of this propensity to recycle and reuse: "It would seem that our industrial society has some characteristics usually considered to typify 'primitive' economies."

Americans tend to think of "recycling" as a relatively modern conceit that has only recently gained broad public acceptance, and whose practical benefits have only just begun to be realized. Among some environmentalists, the need for recycling seems to be viewed as the conceptual outgrowth of a series of propositions that gained wide currency some two decades ago. These are: First, that the United States was potentially on the verge of a new political-cum-spiritual-cum-environmental awakening (a proposition embodied in Charles Reich's *The Greening of America,* 1970); second, that it was about time that something like this occurred, because people were wolfing down the planet's resources far too quickly (a proposition embodied in the 1972 Club of Rome report, *Limits to Growth*); and third, that the new awakening, should it come, ought to be characterized by an ecologically conservative and individually responsible lifestyle (a proposition embodied in E. F. Schumacher's *Small Is Beautiful,* 1973). Recycling has been embraced by some with an almost religious intensity.

This sort of thinking may very well have helped to encourage the emergence of a certain kind of recycling, but recycling itself is probably as old as—indeed, seems to be a fundamental characteristic of—the human species. The archaeological record is crowded with artifacts that display the results of recycling behavior. In ancient times pottery was the equivalent of our own containers made of glass and PET plastic and expanded polystyrene foam, and it was extensively recycled: Broken pottery, in the form of sherds, was frequently ground and used as temper in the manufacture of new pottery. Around the world, masonry structures are often found to harbor stone that had once been used for another purpose, such as the grinding of meal. Building new structures with materials from old ones has always been widespread. Much of medieval Rome, to cite one famous instance, was constructed of marble and other stone scavenged from buildings erected in imperial times; the Colosseum served for centuries essentially as a quarry. In the Industrial Age, cost-conscious manufacturers have become adept at making the most of raw materials—for example, by finding uses for many of the "useless" byproducts of various factory processes. Waste materials discarded in the course of trimming, cutting, scraping, and so on are routinely reintroduced at the start of the manufacturing cycle. And, of course, there still exists a scavenging-and-recycling regime in many Third World countries that is little different, save for the obvious matter of scale, from that employed by our distant ancestors.

As already mentioned, in Mexico City today there are some seventeen thousand garbage-pickers, or *pepenadores*. They systematically pick through the garbage delivered to the capital's sprawling open dumps—it has already been picked through, needless to say, by the more privileged men who push the garbage carts and the even more privileged men who drive the trucks—looking to reclaim cans, bottles, cardboard, scrap metal, broken appliances, paper, plastic sheeting, and meat bones (these last destined for the manufacturers of bouillon cubes and glue). Food debris is consumed by the herds of hogs that the *pepenadores* tend at the dumps. The *pepenadores* are tightly organized by a network of *caciques*, or headmen, and are as politically powerful as Mexico City's transportation workers and those workers who keep the capital's deep-drainage system in working order. (Mexico City is built atop what five hundred years ago was a lake, Lake Texcoco.) The whole *pepenadore* system depends, of course, on the fact that the average daily wage for the typical worker in Mexico City is so appallingly low. The amount of money that can be made by scavenging may be modest in absolute terms, but it is sufficiently great in relative terms to serve as the engine that drives a vast recycling system. Internally the system is also driven by money, by increments of marginal advantage: This is apparent, for example, in the *pepenadores'* pyramidal pecking order, which reserves to certain classes of garbage worker the scavenging rights to Mexico City's wealthier neighborhoods.

In Cairo, where a system similar to Mexico City's is employed, the city's scavenging *zabaline* manage to recycle some 80 percent of what they pick up, down to the filaments in light bulbs. Like other professional scavengers, they tend to specialize: one family will handle mostly plastics, another rags, another glass. Jasper Bouverie, a writer and editor with the magazine *Cairo Today,* described all this recently in the British journal *New Scientist.* Among Bouverie's discoveries, after a visit to a *zabaline* community called Manshiet Nasser:

> There is also a thriving market in Western spirit bottles, such as Johnnie Walker, Gordon's Gin, and Chivas Regal, which fetch about 50 piastres (9p) each. They are sold to a handful of local companies which use them to bottle their own spirits. The companies then use similar labels and give their products names such as Johnnie Darkie, Dordon's Dry Din, and Chivas Renal, in imitation of the original products.

Owing to the efforts of the *zabaline*—who, incidentally, are collecting only a little more than a third of Cairo's garbage—the overall garbage-recycling rate in Cairo is estimated to be about 30 percent, which is more than double the rate achieved in the United States thus far.

A price-and-wage structure in which even the basest material commodities are more valuable than human beings is a fool-proof recipe for a recycling program of prodigious capacity. Though not a recipe most Americans today would be willing to accept, this type of economy, still the rule in much of the world, was a reality for large numbers of people in the United States as recently as the turn of the century. "Rubbish in the streets and in the alleys," observed Rudolph Hering and Samuel Greeley in their 1921 book, *Collection and Disposal of Municipal Refuse,* "is picked over by a class of men who gather anything that has a ready commercial value. The number of men who are thus employed is large; nearly all junk dealers are ready to furnish them with carts and bags. The work is systematized, the men working in definite districts."

Boston's elegant Back Bay neighborhood, once a tidal marsh, owes its existence not only to ample loads of gravel and other kinds of fill but also to the copious amounts of garbage that Bostonians dumped into the site. An etching done by Winslow Homer in 1859 (Figure 9-A) shows scavengers—"chiffoniers," he delicately called them, "pickers up of unconsidered trifles"—combing furiously among the Back Bay garbage dumps. The trade in rags was particularly important, with a class of material known as "thirds and blues" (rags that were third-hand and blue or lighter in color) being a staple of paper-making and of reprocessed or "shoddy" clothes.

America's system of widespread garbage scavenging was largely destroyed by a variety of factors: the nation's expanding and diver-

Figure **9-A.** *Scene on the Back Bay Lands* (1859), by Winslow Homer. The image of impoverished scavengers picking over massive garbage piles, now a standard element in mental pictures of Third World cities, was once a commonplace one in the United States as well.

SOURCE: The Museum of Fine Arts, Houston. Mavis P. and Mary Wilson Kelsey Collection of Winslow Homer Graphics.

sifying economy, which created more and better jobs; the favoritism in behalf of virgin materials that for years was built into such mechanisms as railroad freight rates and the federal tax code (and to some extent still exists); and, eventually, even concerns about the health of those who worked in the dumps. Though dealers in scrap metals and paper have continued to thrive—as does the Salvation Army, which became active in "household salvage" within a few decades of its founding, in 1865—the idea of scavenging as a desirable economic activity for tens of thousands of individuals more or less receded from public favor. An awareness of recycling receded with it.

One exceptional moment occurred during the Second World War, when Americans and Britons on the home front sorted and saved enormous quantities of tin, aluminum, rubber, paper scrap, and other commodities; the government collected all this material, supposedly for use in the war effort. However, in England, according to Jane Bickerstaffe, the technical director of Britain's Industry Council for Packaging and the Environment, much of the material was simply stockpiled—it was too much trouble to move and clean and process —and, unbeknownst to the public, was quietly landfilled when the war was over. In the end, then, the whole endeavor had, perhaps unwittingly, done good mostly for morale. Apparently much the same thing happened to vast quantities of recyclables collected in the United States. The wartime recycling effort here was also plagued by severe gluts, which upset the secondary-materials markets. Suellen Hoy and Michael C. Robinson, in a monograph written for the U.S. Public Works Historical Society, note that the dutiful public offerings of used wastepaper were so overwhelming as to destroy the paper-collection program entirely. A retrospective (November, 1945) War Production Board report observed:

> The glutting of the market early in 1942 and the resultant price-break created substantial repercussions. The charitable and fraternal organizations which had participated so vigorously in the campaign found themselves with accumulations of wastepaper which could not be sold at an amount sufficient to cover the cost of its collection. The public, which had been urged to accumulate minimum lots of 100 pounds before calling a collector, found itself holding quantities of wastepaper which could not be moved on any basis.

Fearing that disenchantment with paper recycling, which was palpable among volunteer organizations, would spread to other recycling efforts, the War Production Board in June, 1942, asked the public to stop saving wastepaper. The lessons of this episode were not, however, widely taken to heart.

Recycling next emerged in the United States during the late 1960s and early 1970s, with the brief effloresence of hundreds of grass-roots "buy-back" centers (where people drop off recyclable products and are paid for them, usually at some modest per-pound rate). These recycling centers were primarily the handiwork of well-intentioned activists wanting to promote environmental responsibility in a world of finite resources, and almost all of them soon ran into difficulties, victims of the fabled vagaries of the secondary-materials markets. The problem was simple: There was just more recyclable material, more "urban ore," as it is sometimes called, than anyone wanted to buy at prices that would keep the recycling centers in business. Some recyclers began paying to have their mounting piles of paper and other commodities dumped in landfills, but secretly, so the public would not lose heart. (One unwritten law among some recyclers seems to be: Don't ever let the public stop participating in a recycling program, no matter how little sense continued participation makes. The fear seems to be that good habits will be broken.) Because about sixty pounds of material in the average automobile at the time was made from recycled mixed-waste paper—the inside door panels, the back of the front seat, the floor of the trunk, and the bulkhead between the trunk and the cabin all contain paperboard—and because a lot of recycled paper is used in wallboard, the big slump in car sales and housing starts during the mid-1970s helped deliver the coup de grace to recycling by eroding demand for the recyclables themselves.

Despite the widespread failure of these grass-roots efforts, the recycling industry was eventually revived, owing in large measure to forces having nothing to do with recycling per se that had been set in motion in the mid-1960s. Few Americans under the age of thirty-five remember the casual abandon with which many of their fellow citizens threw garbage from car windows as they drove—a habit made all the more evident as the new interstate highway system was linked together, drawing millions of motorists onto the roadways.

Nor was a great deal of social stigma attached to highway littering. The authors of a study published during the 1960s by the National Research Council's Highway Research Board, and based on data from twenty-nine states, found that along a typical mile of American highway one could retrieve 710 beer cans, 143 soft-drink cans, 227 glass bottles and jars, 155 pieces of plastic, 352 paper packages, 58 newspapers and magazines, and 1,195 other pieces of paper. The total number of littered objects per mile averaged 3,279. It was Lady Bird Johnson as much as anyone who, campaigning from the White House on behalf of her national beautification program, gave the United States an improved aesthetic sense of self, and helped to change public attitudes toward litter and other forms of roadside ugliness. "Ugliness creates bitterness," Mrs. Johnson told a White House Conference on Natural Beauty in 1965. (The very name of the conference seems to underscore how far we have come, or perhaps fallen, since then.) "Ugliness," she went on, "is an eroding force on the people of our land."

Garbage that is out of place—the definition of litter—has always attracted far more attention than garbage that winds up where it is supposed to, a fact that, as we have seen, is partly responsible for so many of the myths about landfills. Being visible, litter tends to make news, as did the report last year (in the journal *Nature*) by a British zoologist, Tim Benton, who on a 1.5 mile stretch of Ducie Atoll, an uninhabited island in the Pacific Ocean that lies 293 miles away from the nearest inhabited island, found 953 pieces of litter, including 171 bottles, twenty-five shoes, six light bulbs, three cigarette lighters, one football, and an asthma inhaler. All states now publish annual statistical summaries of the litter situation on their highways, and various formal and informal studies have revealed many of litter's enduring characteristics. For example, the rate of littering declines by about half when it is raining. Three litterers out of four are male, and three out of four are under age thirty-five. One early study of litter on the University of Arizona campus, by a student named Judith McKellar, indicated that, bottles and cans aside, there was a critical size factor in litter disposal patterns. Items that were at least four inches in size along at least one dimension were consistently found in trash cans; those items that became litter were almost always smaller than four inches in size along all dimensions. James O'Connell, whose work on discard patterns in the Australian outback was cited in chapter two, has determined that a similar size distinction exists between the garbage found in activity areas used regularly by aborigines (it tends to be smaller than four inches along all dimensions) and the garbage found beyond those areas (it tends to be larger than four inches along at least one dimension). O'Connell has professed to be intrigued by the possibility of a general "McKellar Principle" of littering.

In any event, one eventual response to the litter problem was a proliferation of bottle bills, which typically mandated a five- or ten-cent deposit on certain glass and metal beverage containers, and provided for a return of the deposit when the containers were brought back to the point of purchase. Additionally, beverage distributors were usually required to take back these containers, which could then be sold on the scrap markets or, if reusable, be reused. Besides a reduction in litter—and bottle bills *are* effective in this regard, their financial influence on consumers, thrifty children, and penurious itinerants being a major reason why the "capture rate" of bottles and cans is extremely high—it was anticipated that legislation of this kind would also save energy and natural resources, and reduce the volume of municipal solid waste. Oregon passed the first bottle bill in 1972. Among other things, it banned pull-tab cans (the tabs represented the sort of omnipresent litter that only an outright ban could combat—their archaeological value be damned). Other states soon followed suit.

So far so good. Recycling, however, happens to be one of those endeavors whose potential participants—local governments, retailers, environmentalists, scrap dealers, manufacturers—are divided one from another by complex divergences of interest. In this case, the opponents of bottle bills included supermarkets, which faced the prospect of onerous new demands (sorting and storage) with little to show for it; the manufacturers of one-way beverage containers, whose livelihoods would obviously be diminished; and many of the makers and distributors of beer and soft drinks, who had to collect the empty containers and return them to bottling plants. The nation's biggest brewers of beer and soft drinks were among the most adamantly opposed to bottle bills, and their efforts at resistance, though not sufficient to wipe bottle bills off the face of the earth, had the ironic result of creating a complementary form of recycling.

Bottle bills threatened the competitive forces that were shaping the beer industry's giants. Returnable bottles represented no real hardship to the many local brewers who served a particular city or region; they had been reusing their bottles for years. But returnable bottles were a hardship indeed to companies like Anheuser-Busch, Miller, and Coors: companies that either already served a national market or had plans to do so; that hoped to undermine local brewers; and that depended on being able to send millions of bottles on long-distance, *one-way* voyages of conquest. The big brewers adopted two strategies to combat bottle bills. One was the usual head-on assault by lobbyists and public-relations specialists. The other was more subtle and, in unexpected ways over the years, vastly more productive: the opening of recycling centers all across the country, under the aegis of something called the Beverage Industry Recycling Program (or BIRP). In effect, the brewers were offering states a deal: You get us off the hook on bottle bills—on having to collect and reuse the containers we sent out into the world—and we'll solve the litter problem in our own way. The beverage industry would set up recycling centers which would offer a financial incentive to consumers just as the return of a deposit did, thereby drawing beverage containers out of the environment (just as a bottle bill does). Containers that were returned would then be recycled.

The upshot of all the contentiousness over bottle bills is that through one means or the other (or both) a variety of recycling programs got off the ground almost everywhere. Today ten states have bottle bills of some kind (and such bills are under consideration elsewhere), and just about every state has a BIRP office and recycling businesses that evolved from BIRP. Some of the bottle bills that have been passed contain curious features. For example, owing to remonstrations from the beverage industry, which cited the success of voluntary recycling programs, the Delaware legislature when it passed its bottle bill in 1983 exempted aluminum cans from the bill's provisions. However, because Delaware's farmers were concerned that aluminum cans demanding no deposit might continue to be thrown from passing cars into fields, where the cans could be ground up by farm equipment and then ingested by cows, the beverage industry agreed to establish a $20,000 Bovine Trust Fund, administered by the secretary of agriculture of the state of Delaware, which would reimburse a farmer for any cow (or horse) whose demise was certified by the state veterinarian to have been caused by eating one or more aluminum cans. "It is our understanding," Stephen K. Lambright, a vice president and group executive of Anheuser-Busch, said in a statement in 1991, "that to date, no claims have been made on the fund."

If litter served to jump-start the recycling movement, the aluminum can has kept the engine running. Recyclers deal in newspaper, cardboard, and glass bottles, as well as aluminum cans, but from the beginning aluminum has been by far the most lucrative to recyclers and therefore the most desirable of the recyclable commodities. The official estimate of the aluminum recycling industry is that some 60 percent of all aluminum cans today are recycled. Given the fact that used aluminum cans sell for about $400 to $600 a ton, and that beverage companies use more aluminum in a year than does the U.S. military, the automobile industry, or the aircraft industry, it is not surprising that aluminum has accounted for a disproportionate share of revenues at buy-back centers. In a city the size of Phoenix, Arizona (population: 985,000), according to a Garbage Project analysis done in 1988 for Phoenix's department of public works, the aluminum cans discarded in a typical year would, if sold at market rates, bring in a minimum revenue of $6,372,000. Throughout the 1970s and early 1980s aluminum cans, once a reviled symbol of roadside litter, secured the wobbly infrastructure of those recyclers who depended on materials collected from individuals. (The aluminum cans collected by the Garbage Project in the course of its investigations all eventually find their way into the recycling stream, either through Boy Scouts, to whom they have been given, or via homeless people, who may climb over the fence and take them.) Not coincidentally, during this same period of time roadside litter was reduced considerably and the issue lost some of its prominence, although it is still getting attention in some states. In any event, by the late 1980s concern over litter had helped to ensure that the idea of recycling was fairly well established.

Today, the full corpus of laws within each state bearing on all aspects of recycling, enacted piecemeal over many years, is something of a hodge-podge, and the situation becomes more confusing still if one compares states with one another. In 1990 more than 140 laws that somehow bear on recycling were passed in thirty-eight states; at any given moment hundreds more are pending. Thirty-three states and hundreds of cities have recycling programs in place that are described as somehow "comprehensive." Some programs are voluntary and some are mandatory. Some programs rely on buy-back centers, others on curbside separation, others on separation by human hands at materials-recovery facilities (MRFs—pronounced "merfs"), still others on a combination of approaches. Whatever the specific procedures, however, the fact is that the job of *collecting* household recyclables is a challenge that is now beginning to be met. As the recycling experience of the Second World War and of the early 1970s suggests, however, the challenge of collection is not the only one.

Long before the recent spate of recycling laws; long before the creation, in October of 1970, of the Environmental Protection Agency; and long before the celebration, on April 22 of that same year, of the first Earth Day, that chronometric marker for the environmental movement as a whole—before all of this, there existed a breed of entrepreneur whose chief aim in business was to collect material objects that had been thrown away and sell them to people who had a use for them. These entrepreneurs, referred to earlier, are the scrap-metal and secondary-materials dealers, and most of them belong today to an organization called the Institute of Scrap Recycling Industries (ISRI), which was created in 1987 through the merger of the Institute of Scrap Iron and Steel (ISIS) and the National Association of Recycling Industries (NARI). The secondary-materials business is largely a family business (albeit with some big corporate participants thrown in, such as Commercial Metals, in Dallas, and Proler International, in Houston), and many of the families have operated their scrap yards—the trading pits of the commodities market in recycled materials—for three or four generations. The yards are not much to look at, to say the least. Frequently located on the edge of town or down on the waterfront, and today often shielded from nearby interstate highways by a wall of sturdy fir trees (thanks once again to the inspiration of Lady Bird) or high barriers of wood or metal, the yards groan under their mountains of baled newspaper and cardboard, their railroad cars filled with crushed cans, their rusting stacks of flattened automobiles and flattened white durables.

Visit the annual convention of the Institute of Scrap Recycling Industries and you will see the membership scrubbed for public display. Attendance at the convention typically numbers three thousand, which means that there will be nearly that number of gold nugget rings, gold neckchains, and vinyl belts. There will also be that number of beepers and cordless phones. But the scrap industry still retains many of the elements portrayed in the 1950 movie *Born Yesterday,* with Broderick Crawford and Judy Holliday: the risk, the uncertainty, the brash seat-of-the-pants maneuvering.

ISRI members are businessmen who possess arcane forms of knowledge, and they do their job with great efficiency. These are people who, when it comes, say, to buying or selling paper, know exactly what is meant by terms like "used brown kraft," "special news/de-ink quality," "mill wrappers," "flyleaf shavings," and "sorted colored ledger"—and what the price history has been for each grade during the past year, and what the price is likely to be tomorrow in Trenton or Taipei, and where they can find freighters for their wares that might otherwise be returning home empty but can backhaul to everyone's benefit, and so might be had for a song. They know how to use guillotine shears and shredders to rip cars into pieces, and how to use baling presses to compress cars into cubes. Because they do, the materials in just about all of America's junked cars are almost completely recycled. In 1990, ISRI members recycled some nine million automobiles—only slightly fewer than the number of new cars registered in the United States that year.

Collectively, the members of ISRI process and sell some sixty million tons of ferrous metals annually, along with seven million tons of nonferrous metals, and thirty million tons of waste paper, glass, and plastic—almost 100 million tons of material in all. The size of their operation in the aggregate dwarfs that of all recycling programs run by governments at the city, county, and state level taken together. The high-minded public servants around the country pressing hard for recycling legislation and municipal recycling efforts sometimes spoke, when they were getting into the business, as if they were colonizing virgin territory; in fact, as these officials have come to realize, the territory is already populated, and the natives are resourceful.

The most important lesson that the natives have to teach us, and the one most frequently lost sight of, is the importance of market forces as the power plant of recycling. Money powers recycling as surely as the sun's energy powers the winds; absent the money, and recycling lies becalmed. The popular image of what constitutes recycling—separating one's garbage into various categories, leaving it neatly sorted at curbside, and seeing it carted off by industrious sanitation workers—does not really constitute recycling at all. It constitutes *sorting* and *collecting*. Recycling has not occurred until the loop is closed: that is, until someone buys (or gets paid to take) the sorted materials, manufactures them into something else, and sells that something back to the public. The unbudgeable primacy of this economic fact is why the members of the Institute of Scrap Recycling Industries sometimes look with skepticism at those state and municipal recycling programs whose first concern has been not with expanding markets for recyclable materials but instead with increasing the supply of recyclable materials—that is, with simply getting garbage "source-separated" by households or sorted by recycling centers.

The point was driven home one day about a decade ago when a Garbage Project representative was invited to give a talk in Calgary, Alberta, at the annual convention of ISRI's Canadian counterpart, the Canadian Association of Recycling Industries. During a conversation with one of the CARI members the night before the talk the Garbage Project representative began describing the city of Tucson's voluntary newspaper-recycling program. He explained that citizens were asked to keep their newspapers separate from their other garbage and on collection day to leave the newspapers out in paper bags, or in neatly tied bundles. The newspapers would then be carted off and recycled. Another blow struck for resource conservation.

The CARI member's expression was pained, and his subsequent remarks achieved a pitiless, withering tone not normally associated with Canadians. "You're telling me how well our competition is doing," he said. "The ones who are subsidized by the taxpayer when they try to take away our livelihood. Don't you understand? There never has been a shortage of newspapers to recycle. The shortage is in demand. Markets fill up just like landfills. There are just so many car panels and cereal boxes that need to be made. I suppose you believe that GM is going to say, 'Hey, great! Here's a bunch more newspapers we can recycle into door panels. Let's make some more cars!' The more the city of Tucson recycles, the less I do." Needless to say, the Garbage Project representative's speech was amended that night in one or two particulars.

As noted, the importance of recycling as a solid-waste-management strategy is now broadly accepted in the world beyond that of the scrap dealers and factory managers who have always seen its worth. The Environmental Protection Agency has set a goal for the year 1992: it hopes that by year's end some 25 percent of America's municipal solid waste will be dealt with through recycling. That deadline won't be met—as of this writing, only about 15 percent of all American municipal solid waste is being recycled. There does seem to be wide public support for recycling, built on the grounds—the unassailable grounds—that recycling is a good thing. But, as history shows, recycling gets done not because it is a good thing; it gets done if it is a profitable thing, and profitability in this case depends primarily on the demand for recyclable materials.

How much demand is there? As one would expect, the situation with respect to aluminum is rosy: Experts agree that demand for recyclable aluminum is virtually unlimited. Making new aluminum from virgin materials is costly; it requires twenty times more energy than does the processing of aluminum cans back into aluminum ingots, and the cost of bauxite ore is high. All the aluminum companies have overcome their initial resistance and become committed recyclers. Perhaps as many as 55 billion aluminum cans were recycled in 1990, and the figure is likely to grow significantly year after year. Altogether, about 30 percent of all the aluminum produced in America is reused.

Used glass is broken up into cullet and added as temper to strengthen the bond in new glass—a technique, as we have seen, that has echoes in the manufacture of ancient pottery. The demand for cullet would seem to be limited only by how much new glass is made. However, there do exist some constraints. Glass must be sorted by color (so that new glass made from old glass doesn't cloud). It is also extremely heavy and therefore expensive to move; recyclers pay out a lot more money for transportation than they do for the glass itself. Finally, the recycling of glass is not being driven by any concerns of cost or resource scarcity. No glass manufacturer is seriously concerned that the world is running dangerously short of the raw material glass is made from (sand), or that prices will go through the roof. Nevertheless, about one glass bottle out of five gets recycled these days.

Plastic, too, has begun to show promise. The preference shown by consumers for soft drinks sold in PET plastic bottles (the invention, by the way, of the late Nathaniel C. Wyeth, a son of the illustrator N. C. Wyeth and the brother of Andrew, and a member of the Society of the Plastic Industry's Hall of Fame) and for water, milk, and orange juice sold in HDPE plastic bottles finally created a situation where collection programs could conveniently capture large quantities of plastic of a homogeneous type, a prerequisite for efficient recycling. (Mixed plastics are difficult to reprocess into anything people will pay much for.)

The most successful company that works with recycled plastic has been Wellman, Inc., based in Shrewsbury, New Jersey, which in 1990 consumed more than half of all the PET plastic collected for recycling. (This in turn represented about 30 percent of the 35 billion PET bottles sold that year.) Wellman first ships the plastic to processing plants in Johnsonville, South Carolina, and Mullagh, Ireland; each PET bottle is cleaned of its aluminum cap and paper label, and separated from the black HDPE base cup. The HDPE is sold in flake form to be made into plastic handles for irons and other tools. The PET itself is turned into either fiberfill stuffing for sleeping bags and ski jackets or into polyester carpet fiber. Other companies are getting into the act. In 1990, both Pepsi Cola and Coca Cola sought an exemption from the Food and Drug Administration's regulation prohibiting recycled materials to be used as food and drink packaging, and last year the exemption was granted for recycled PET. Companies are also being pressed toward various kinds of plastic recycling by cities that have banned certain plastic products or threatened to ban them (a tactic known as "greenmail"). Overall, the situation with respect to plastic seems guardedly optimistic; prices for some plastics are very high, and they, like aluminum, could experience an image enhancement as a result. As yet, though, only a tiny fraction of the plastic in the waste stream, about 2 percent, is recycled.

While demand seems generally to exceed supply for aluminum, and to match or exceed it for glass and plastic—but not always or everywhere—it is important to bear in mind that these commodities amount to only a small fraction of a typical landfill's content (a total of only about 18 percent by volume). Even if the percentages of these commodities that are recycled increase dramatically, the result will not be yawning chasms in the local landfill. Increased recycling of paper (which accounts for 40 percent or more of landfill volume), on the other hand, could make a considerable difference, but it is with respect to paper, particularly newsprint, that the economics of recycling have yet to work out. Most recycled newsprint in this country is either manufactured into fresh newsprint, boxboard (if a box—say, a cereal box—is gray on the inside, it has been made from recycled paper), wallboard, insulation, and automobile interiors, or is shipped to sundry manufacturers overseas. In recent years two new markets for recycled newsprint and phone books have opened up: in the making of toilet paper (the Forest Green brand, for example) and the making of "shred bed" (which is a substitute in cattle stalls for straw). Many of the end uses for recyclable paper are now at or near saturation, and as a result certain kinds of paper that have been collected by recyclers sometimes can't be *given* away.

Consider what has happened to newspapers. During the late 1980s laws began coming into effect in the Northeast that mandated the creation of state or local recycling programs. In New Jersey, legislation went into force in 1987 requiring that every community in the state begin to set aside, for the purpose of recycling, any three commonly discarded commodities: for example, aluminum cans, PET bottles, and newspapers. At the time, New Jersey was already collecting 50 percent of its newspapers (the national average today is about 33 percent), but within a few months of the law's taking effect the proportion of all newspapers being collected had soared to 62 percent, and the price of newsprint had fallen from $45 per ton to *minus* $25 per ton—that is, recyclers had to pay someone to take the newspapers away (possibly to a warehouse; possibly to a landfill).* What happened in New Jersey happened to towns and cities throughout the Northeast. As mandated collection programs continued coming on line, the ripple effects began to spread further afield. Even if the economics were not sometimes disastrous, there simply are not enough mills in the United States to process all of the paper being collected here, and some paper companies have proved gun-shy about making the $500 million investment that each new paper-recycling mill requires, or even the $40 to $60 million investment that the retrofitting of existing facilities would entail. Retrofitting also takes a lot of time: four to six years.

One further result of the present glut has been that hundreds of thousands of tons of American paper are now essentially being dumped on the European and Asian markets for next to nothing, causing paper prices there to fall and disrupting—in some cases, destroying—finely tuned European and Asian recycling programs. In Holland, for example, the price of surplus paper dropped from eight cents a kilo to under a penny a kilo in the course of 1990 alone. Soccer teams and other local groups that once relied on paper drives for their funding suddenly had to scramble for alternative means of support. In Japan, the influx of cheap American newsprint (with very little recycled content, Japanese buyers are pleased to note) has reduced demand for the homemade product (with its very high recycled content), and for a time threatened the livelihood of the "toilet-paper-people," who go from door to door exchanging toilet paper for old newspapers.

Recycling is a necessary component of a sound solid-waste-management program. Properly conceived and executed, a recycling program can make good economic sense, can help save natural resources, can help reduce pollution, and can divert some tributaries

* Although oversupply caused trouble enough, there was an additional problem in New Jersey. Before the 1987 law went into effect, paper collected for recycling was kept separate by category: old newspaper with old newspaper, office paper with office paper, and so on. The new law did not require such separation, and the paper that began coming on line was collected as "mixed"—a cheap grade, owing to the need to have it sorted. The influx of mixed paper made the price decline all the worse.

of the solid-waste stream away from landfills. These are all essential goals. There is no reason, however, for recycling to become an individual or social obsession. Indeed, when recycling does become an obsession in a society, it is sometimes a sign that important aspects of that society have gone seriously awry; as they have, for example, in the Soviet Union, where the scarcity of even the most basic consumer goods has driven the populace to the most desperate frenzies of recycling imaginable. Still, without becoming obsessed by recycling there are useful, pragmatic steps that Americans can and should take to bolster the recycling enterprise—in particular by fostering demand for recycled materials. These steps include buying consumer goods that have truly been made from recycled materials (beware of misleading claims on the packaging) and buying consumer goods that, once discarded, will have the most resale value for the recyclers. (We will return to this subject in chapter eleven.)

In the meantime it must be remembered that while recycling is one valuable way of coping with America's—or any society's—solid waste, it is by no means a panacea. Yes, from a narrow, technical perspective almost anything that one might find in municipal solid waste *could* be thought of as being somehow recyclable or reusable; the problem is finding significant outlets for such recycled or reused products that also make economic, political, environmental, and psychological sense.

For one thing, it is not farfetched to think that recycling may one day be met with antagonism by its erstwhile middle-class allies. Despite the virtuous public image that recycling possesses when considered in the abstract, in the real world recycling could find its reputation tarnished. There have already been reports, for example, of inroads into the recycling business by organized crime. On a more mundane level, garbage-sorting centers and recycling centers, like any public-works projects, are increasingly becoming objects of NIMBY-type opposition. Most recycling centers and plants are nothing more than enclosed spaces where presorted cans, bottles, and newspapers are temporarily stored or, at worst, where mixed recyclables passing by on conveyer belts are separated by human hands. They do not belch noxious fumes. The work being done inside them may very well be God's. But, BUT, they bear the unholy taint of garbage. And don't forget all those trucks coming and going all day long. As the collection and sorting of garbage for recycling become a growing and regular part of our lives, so will protests against conducting these activities anyplace nearby.

The large-scale composting of municipal solid waste is touted by many as one way of recycling the 5 to 20 percent of household garbage that consists of yard waste and food waste. (The yard-waste volume varies considerably by region.) But composting also confronts NIMBY problems, and environmental concerns of other kinds as well. Composting is an ancient practice—there is evidence that composting pits were in use at Knossos, in Crete, some four thousand years ago—and in theory composting seems compelling and attractive. Yard waste has been banned from landfills in more than ten states precisely in order to encourage small-scale composting at home, and a number of companies now sell small plastic "green cones" for this purpose. Large-scale composting of municipal solid waste is something of a different proposition. The enormous volume of rich humus that results from large-scale composting has a variety of commercial uses, and the Europeans began resorting to a significant amount of composting years ago. But composting is expensive. Composting yards are also big, and they can smell; siting them may not prove to be as difficult as siting landfills, but doing so will still take a lot of work. Moreover, if precautions are not taken to prevent certain kinds of biodegradable garbage from joining the compost piles, the compost can become tainted with hazardous elements, such as the heavy metals in inks and pigments. Yard waste may contain traces of pesticides and herbicides. Composting is only just getting under way in the United States, and there are as yet fewer than a hundred composting plants planned or in operation, most of them small. A lot of thought is being given by composting proponents to ways of dampening potential opposition (such as making sure that the composting piles are physically enclosed, to contain "fugitive odors"). But this industry, if such it becomes, is starting out with some handicaps.

Recycling of other kinds exacts an environmental price. The reuse of paper, for example, involves processes that generate a considerable amount of hazardous waste. In order to recycle newspapers, magazines, and, indeed, any printed paper, the paper must first be de-inked. At the end of the de-inking process one is left with essentially two products: on the one hand, de-inked fiber that will be turned into new paper; and on the other, a large quantity of toxic sludge. The recycling of iron and steel, of aluminum, and of plastics, for their part, also result in the production of various kinds of toxic waste and in air emissions that may be hazardous. A 1988 U.S. Office of Technology Assessment report on solid waste observed bluntly of recycling that "it is usually not clear whether secondary manufacturing produces less pollution per ton of material processed than primary manufacturing."

Another vexing reality that communities must confront is that recycling can be expensive. A myth was once abroad that recycling was not only an environmentally sound garbage-disposal option but also a potential money-maker or at least money-saver. That this was going to be the case was at least implicitly the notion that lay behind the various "zero-net-cost" recycling schemes many communities adopted. The idea here is that when a city (for example) considers bids from independent recyclers for handling its recycling program, the amount the city finally agrees to pay per ton must be no greater than the cheapest available disposal method other than recycling (which is usually landfilling, the high cost of which in some places is what makes recycling attractive to begin with). The assumption, of course, was that the recycling agency would earn enough from sales of recyclables to more than offset the difference, if any, between city fees and the actual cost of operations. This frequently has turned out not to be the case for private recyclers of household-level commodities, and the same economic realities that bedevil them, of course, also bedevil recycling programs operated by communities themselves. From the start, recyclers have been beset by slumps in commodities prices. The cost of collection programs is high to begin with —think of the capital investment required for new kinds of trucks. There have also been unexpected problems. One major glitch collection programs have faced: the inability of many consumers to sort their recyclables properly prior to pickup, resulting in "contaminated" deliveries that may be rejected by buyers or must be sold for reduced prices. Most cities have had to set up costly labor-intensive or mechanical sorting operations to sort once more the garbage that households have already sorted.

The problem of improper sorting by households is pervasive. The

Garbage Project last year sampled the sorting behavior of twenty randomly selected households in a middle-income neighborhood in Tucson. Under the rules of the local curbside recycling program, residents were to separate out all recyclables and place them in a special blue bag. The rest of the garbage was to go into the traditional garbage can. What is and is not recyclable depends, of course, on what the community has decided to recycle, and all participants in the recycling program were given clear definitions of what should and should not go into the blue bag. The results of the Garbage Project's survey of the contents of the twenty blue bags and the garbage cans that went with them were not really unexpected. Taken together, the discards in the blue bags weighed 318 pounds, but fifty of those pounds were taken up by nonrecyclables. Every household made mistakes, either by contaminating the recyclable bag or throwing recyclables into the garbage cans. Indeed, fully half of all the aluminum cans thrown away were found not in the recyclable bag but in the garbage can.

This is why materials-recovery facilities (MRFs) are needed. Once again the American consumer has proved capable of dashing the fondest of hopes. The inevitable consequence of this and other developments is that, far from being a gold mine, recycling will be a procedure—a worthwhile procedure—for which communities must pay considerable sums, often unexpectedly, perhaps consoling themselves with the recognition that resources have been conserved and that some garbage has been kept out of landfills and incinerators.

A further reality that will become apparent with time is that some significant elements of the solid-waste stream that are without question recyclable will prove resistant, for a variety of reasons, to all attempts to recycle them. Rubber tires have so far proved to be a case in point. Tires are every landfill manager's nightmare. They possess a peculiar property: Bury them in a landfill and over time they will slowly rise and eventually emerge onto the surface, as if all the raisins in a loaf of bread had ascended to the top. (One explanation given for this phenomenon is that landfill compactors initially compress the hollow, newly arrived tires, but that over time the tires expand back into the original shape; the act of expansion gradually takes them upwards because the garbage above them is always less compact than the garbage below. An alternative explanation—

which also explains how rocks rise to the surface of a pasture—attributes the phenomenon to temperature fluctuations that cause the tire to expand and contract, with small particles drifting into the tiny void that forms under the tire after each contraction; slowly but surely the tire works its way to the surface.) Periodically landfill operators skim the landfill surface with a special vehicle that picks up the latest crop; this is the source of those large tire islands that one sees alongside most landfills, and that every so often ignite uncontrollably and blacken the skies for weeks. In theory, tires would seem an ideal candidate for recycling. There are lots of them—200 million are thrown away every year. They are relatively homogeneous. They even, as we have seen, eerily separate themselves from all the other garbage. And yet nothing that has yet been tried—not using them to make road surfaces or airport runways, not burning them (along with coal) as fuel, not using them for artificial reefs—has made anyone terribly excited (or made anyone much money).

Finally, to repeat, recycling can be a surpassingly fragile enterprise. There are many variables, and their configuration from place to place must dictate strategy and tactics. Some kinds of recycling, such as of aluminum cans, probably make sense everywhere. Other kinds, such as of newsprint, may not. Homogeneity of materials may be a necessity when it comes to recycling, but homogeneity of policy across geographical boundaries ought not to be the watchword with respect to how much of what kinds of garbage America's communities should be recycling. The key is to maintain a tautness between supply and demand, a task that is not always easy and, frankly, not always possible. It may become increasingly difficult as the many collection programs that have been enacted into law begin to take hold, and the volume of recyclables on the market suddenly doubles or triples. What may seem like "success" in the eyes of those who run local recycling programs—an outpouring of public cooperation, an Everest of sorted trash—can at times spell failure for the system as a whole, dooming truckloads of recyclables to be dumped into landfills (as some are even now), driving local programs bankrupt, and, depending on the degree of overabundance of this or that, and its effect on prices, even threatening the health of scrap dealers. Too many communities around the country are now reading headlines like this one from a Boston-area newspaper, *The Enterprise* (Brockton): "RECYCLING WORKING TOO WELL; Industry Can't Handle Glut of Materials." Too many are now reading headlines like this one from *The New York Times:* "Our Towns: When Recycling Means Too Much of a Good Thing." Or this one from *Waste Age:* "Recyclers Brace for Office Paper Oversupply." It would be an ironic consequence indeed if recycling were undermined by the best of intentions.

The messages to recycling activists: Pay attention to those market factors. Make sure that people in local communities understand the sometimes fickle dynamics of the recycling process. And make sure they understand that recycling has not happened until the loop has been closed.

CHAPTER 10

LIFE-STYLE OVERRIDE

Source reduction is, on the face of it, perhaps the most appealing of all the possible approaches to solid-waste management. The Environmental Protection Agency and most environmental groups rank it first in their hierarchy of solutions to the problem of garbage disposal, followed by recycling, incineration, and landfilling (in that order). The idea behind source reduction is simple: The solid-waste stream would become at once smaller and safer if we could find ways to minimize the amount of material used in products, extend the useful life of products, and minimize the volume of toxic substances used in products. Source reduction is to garbage what preventive medicine is to health—a means of avoiding trouble before it happens. Source reduction's ideological taproot reaches back to the mid-1960s and the publication of Vance Packard's best-selling *The Wastemakers* and *The Status Seekers,* which railed against America's culture (as he saw it) of planned obsolescence, conspicuous consumption, and industrial manipulation. Packard tended to see the situation in terms of a conspiracy:

> As businessmen caught a glimpse of the potentialities inherent in endlessly expanding the wants of people under consumerism, forced draft or otherwise, many began to see blue skies. . . . What was needed was strategies that would make Americans in large numbers into voracious, wasteful, compulsive consumers —and strategies that would provide products assuring such wastefulness. Even where wastefulness was not involved, additional strategies were needed that would induce the public to consume at ever-higher levels.

By now Packard's bill of indictment seems all too familiar, and there is a righteous, hectoring tone to the book that is off-putting. Nonetheless, source reduction would seem to be the smartest means of reducing the volume of garbage we produce, and the idea has found many adherents not only in various parts of the environmental movement but also in local governments. Its popularity is plainly evident in the rash of product bans or product-disposal bans that have been proposed or voted into effect in towns and cities across the country: bans on polystyrene products made with chlorofluorocarbons (CFCs), bans on polystyrene products altogether, bans on all plastic food packaging, bans on all nonbiodegradable packaging. At least one hundred statewide product bans or product-disposal bans were in place in twenty-nine states and the District of Columbia by the end of 1990.

For all this activity, however, it is not always clear that source reduction pursued in such a manner will have very much impact on the solid-waste stream or even that it will in all cases end up doing more good than harm. The issue of planned obsolescence, to bring up just one example, is not as straightforward as it might seem. First, "obsolescence" per se is not always, or even usually, the reason that people get rid of cars, washing machines, and other durable goods, and buy new ones; rather, people are typically trading up for new conveniences, styles, and colors. Second, the replaced durables are not just thrown away; as we have seen, the castoffs of the good life trickle down quickly to the less fortunate. Third, in a society characterized by continual technological improvement, obsolescence can be advantageous. We are fortunate, for example, that all those gas-guzzling cars of the 1960s and '70s are not still in the pink of health and ruling the roads. The obsolescence of material culture is at once inevitable and essential. This is not meant to serve as an excuse for poor engineering and design; indeed, it often happens that tangible engineering and design improvements are a cause of obsolescence. This much-maligned concept needs to be seen in perspective.

To date, by far the greatest amount of attention paid to source reduction has been lavished on packaging of various kinds, whether the packaging be made of paper or plastic, and whether it is designed to hold food or to hold more durable consumer products. It seems to be taken for granted by many that packaging is inherently wasteful; and, one must concede, some packaging, even a lot of packaging, *is* excessive. But it is also true that consumer industries have strong economic incentives to make products as compact and as light as possible, with the least amount of resources; and, as we have seen, they have responded to those incentives.

The Garbage Project recently looked again at its data for the years 1978 through 1988, with an eye to plotting the story of packaging during the past decade or so (see Figure 10-A). What emerges is that in per-capita terms the amount of material in municipal solid waste that can be classified as packaging has experienced a gradual but real decline. The decline has occurred in packaging made of virtually every kind of material. The amount of paper is down. Aluminum has declined precipitously, thanks to recycling programs. Glass is down because of recycling, light-weighting, and the major switch by beverage bottlers from glass to plastic. Plastic packaging, despite that switch, and owing largely to continuing efforts at light-weighting, has nevertheless held steady in its contribution to solid waste.

The fact remains, moreover, that the effect of modern packaging on the overall size of the solid-waste stream is often not to turn it into a torrent but rather to bring significant parts of it under control. This conclusion can be illustrated by several Garbage Project studies, most notably a comprehensive comparison of household solid waste from Mexico and the United States. The comparison was based on an analysis of pickups of fresh garbage from 1,084 households in a variety of locales in Mexico City (ranging from the desperately poor Las Trancas neighborhood, in the Azcapotzalco District, to the wealthy Lomas de Chapultepec enclave) and from 966 low-, middle-, and upper-income U.S. households in Tucson, Milwaukee,

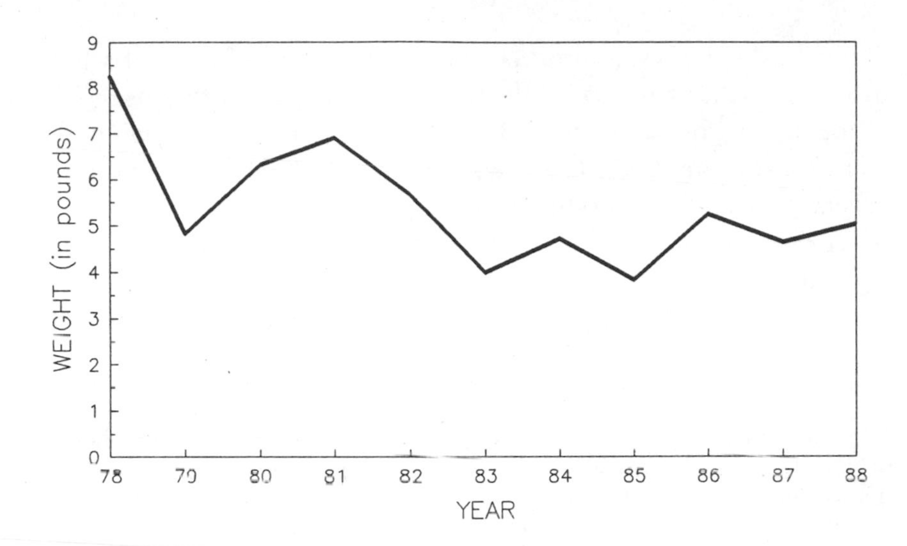

Figure **10-A.** Packaging remains ubiquitous and accounts for a large proportion of all municipal solid waste, but on a per-capita basis Americans in recent years have actually been using somewhat less of it. The graph here shows the amount of packaging found in an average household pickup (biweekly) in Tucson.
SOURCE: The Garbage Project

and Marin County. The differences between Mexican and American garbage are stark and surprising, even after correcting for family size, and the role of packaging goes to the heart of them (see Figure 10-B). In the United States, the skillful packaging of food products cuts down markedly on the wastage of foods. Packaging and the development of a modern, corporate-driven food industry are among the most important reasons why U.S. households, on average, produce fully a third less garbage than do households in Mexico City, where a higher percentage of food is bought fresh, and a larger volume of garbage inevitably results.

To illustrate: If someone in the United States buys any frozen vegetable, it comes in a relatively thin paper box and is devoid of shells, husks, leaves, and other nonedible matter—for all of which, by the way, the processor of the packaged food finds other uses, such as the making of animal feed. The total weight of the packaging may come to, say, an ounce. If one wishes to prepare the same amount of vegetable from fresh produce, however, the shells, husks, leaves, and other organic matter that must be discarded or composted may be equivalent in weight or volume to that of the edible portion. It is worth noting that, thanks again in large measure to packaging, the

Figure **10-B.** The paradox of packaging: The average American household relies on far more packaging, particularly food packaging, than the average Mexican household, but produces far less garbage overall. The data here reflect average daily pickups from households surveyed in the early 1980s.

SOURCE: The Garbage Project

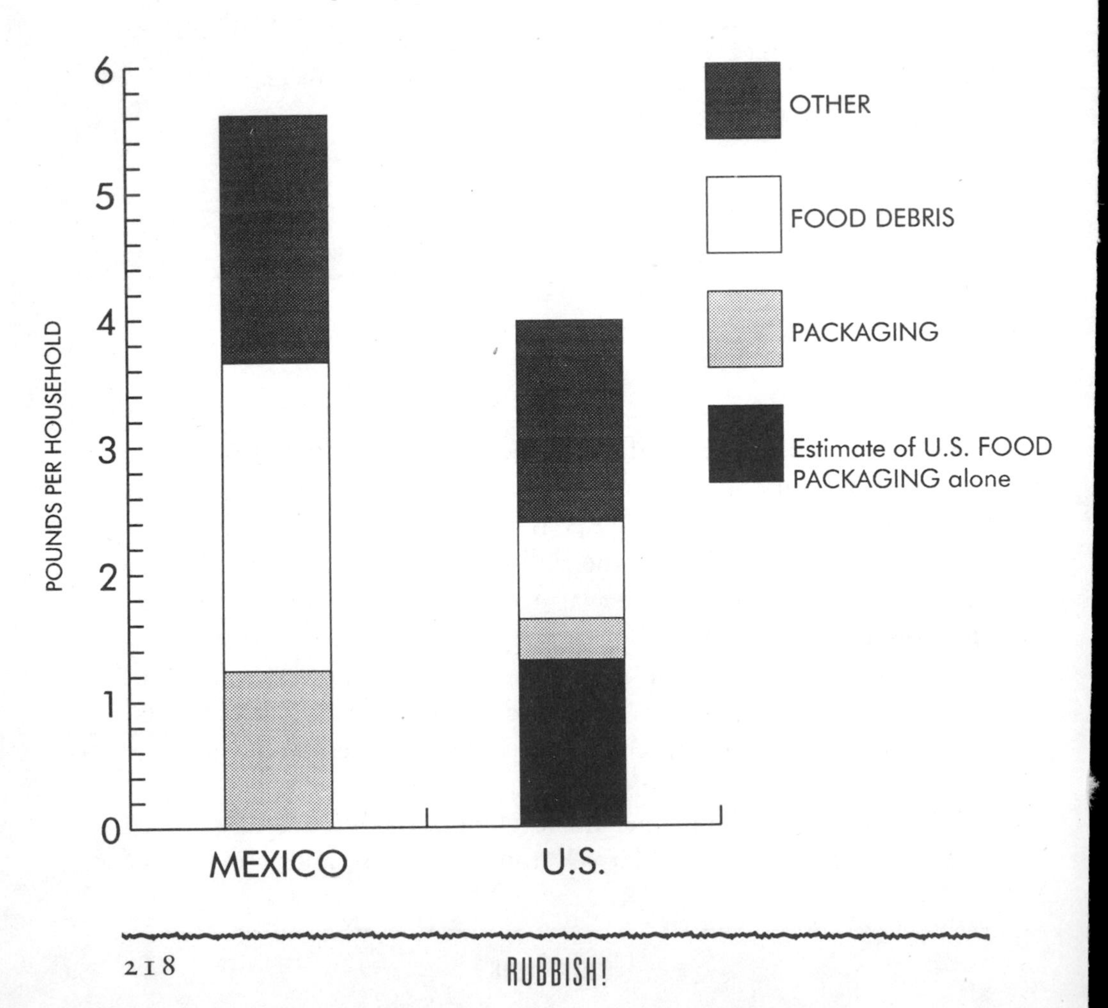

rate of food *waste* in U.S. garbage—as regrettably high as it is—is only about half the rate in Mexico City. Although consumers in the United States waste edible fresh produce at a rate ten times greater than Mexicans do, their overall record on waste is better because of their far heavier reliance on processed, packaged foods, which is the food category that generates by far the least amount of waste.

The point is that, when it comes to thinking about household garbage, something like packaging—whatever kind of packaging it is—must not be looked at in isolation. Although packaging is almost always destined for a quick trip to the garbage can, its sudden demise and all-too-obvious contribution to municipal solid waste should not obscure the fact that it serves larger purposes—purposes that involve such things as efficient resource management, product protection, the prevention of tampering, and maintaining public health (as well as, of course, turning a profit).

Health concerns, for example, are by no means negligible. That vast array of wrappers and boxes at fast-food restaurants, which is the object of so much disparagement, fulfills a role other than mere ease of carry-out. It keeps food safe. Fast-food restaurants and other establishments (such as schools and nursing homes) that rely heavily on throwaway containers disseminate far fewer bacteria and viruses through their disposables—some 50 percent fewer—than sitdown restaurants do through their glassware, silverware, and ceramic plates. The health considerations involved in fast-food packaging at least deserve a moment's thought, which they rarely get. Hospitals, similarly, are unremitting volcanoes of disposable products, many of them plastic; almost all of these products are swaddled in layers of antiseptic packaging. While it may be that hospitals generate so much casual waste because they are being manipulated by industry, or are pandering to it, or because they find it easy and convenient to be wasteful, it is also true that single-use products and packages protect patients and the people who care for them.

These examples are not meant to suggest that improvements in packaging cannot be (or, indeed, are not being) made. They *are* meant to suggest that source reduction is not necessarily simple and straightforward. Furthermore, it should be apparent that what source reduction can achieve, in terms of reducing the amount of garbage we generate, is limited. There are, to be sure, many source-reduction measures for which an argument can sensibly be advanced. For example, it surely makes sense to ban products that are made with chlorofluorocarbons (CFCs), on the grounds that these substances contribute to ozone depletion. (Congress has already enacted legislation that bans the venting of CFCs and provides for the capture, upon disposal, of CFCs used as a refrigerant in appliances.) It may be worthwhile to discourage some mass-use items that are made of mixed materials and consequently are difficult to recycle. But the beneficial results of proposed and existing bans will not include the saving of much landfill space. The banned materials will have to be replaced with other materials, which will still get thrown away.

For source reduction to make a truly eye-catching dent in garbage generation one needs to resort to ideas whose implementation seems improbable—ideas like the ones presented by two professors in a 1978 grant proposal submitted to the National Science Foundation. They proposed to investigate whether manufacturers should be encouraged to initiate development of a form of packaging that would be edible. They also envisioned that homes might one day be equipped with special chutes and hoses that delivery trucks would use to spill or squeeze such items as cereal, peanut butter, and toothpaste into America's kitchens and pantries. The National Science Foundation decided not to look into these ideas further, at least for the time being.

For the most part, the banning of products, save under exceptional circumstances, is, in the end, a not very useful enterprise. The positive achievements are often, at best, symbolic; and, as in the case of food packaging, the downside in terms of higher garbage-generation rates can sometimes prove to be considerable. It should be noted, incidentally, that skepticism about some kinds of source reduction can be heard even within the environmental community. As some have begun to point out, product bans pose a threat to the recycling of certain prevalent kinds of materials, such as polystyrene foam and some plastics, whose futures as recyclables could be bright. In sum, then, there is cause for thinking long and hard before plunging ahead with drastic steps. The appeal of source reduction is undeniable and its potential is real. But, as with the song of the Sirens, one must approach with caution.

Perhaps no corporation in America has so visibly occupied front-line trenches in the source-reduction wars as McDonald's, and its fortunes during the past decade and a half help to illustrate some of the difficulties—and the ironies—of waging war on specific products. The lessons are made all the more pointed when one considers that the corporate philosophy of McDonald's, for all the company's size, is not based on the idea of economic pillage, social exploitation, and environmental rape. McDonald's, as even environmentalists and those who scorn fast food tend to concede, is a commendable corporate citizen—a company that listens to the public and listens to criticism, and tries to respond constructively if only because that makes good business sense.

During the early 1970s—an epoch so long ago in fast-food time that all the signs outside McDonald's restaurants displayed the actual figure for the number of hamburgers sold thus far (as opposed to the weary "billions and billions sold" that one often sees today)—McDonald's and similar establishments came under attack from those who were concerned about the number of trees cut down to make the paper that went into the wrappers that covered the food that satisfied America's burgerlust. Partly as a result of these pressures, and partly in order to keep food warmer and make its packaging less droopy and greasy, McDonald's in 1976 shifted away from paper containers for its hamburgers and some other foods and started using foam clamshells instead.

Little did McDonald's realize at the time that it had simply traded one convenient target for another. The opponents of foam are every bit as vocal as the friends of the trees, though the sources of their agenda are somewhat more varied. Among the most persistent critics of McDonald's has been an organization called the Citizens Clearinghouse for Hazardous Waste, based in Arlington, Virginia; its McToxics campaign has been directed not only at the use of chlorofluorocarbons in the creation of polystyrene foam, but also at the sheer waste that they argue the discarded foam represents, and the fact that foam doesn't biodegrade in landfills. The Citizens Clearinghouse has not wanted for powerful partners-in-arms, and the very visibility of fast-food packaging—combined with the widespread public misperceptions as to its actual volume—ensured that the issue would receive sustained attention. In local communities across the country—first Berkeley, then Portland, Oregon, and Suffolk County, on Long Island—steps were taken to ban polystyrene foam. In 1987 McDonald's sought to dampen the controversy by announcing that its suppliers would stop using CFC-11 and CFC-12 in their polystyrene and switch to a variant form of blowing agent that depletes 95 percent less ozone. Protests continued. Schoolchildren, perhaps with the quiet help of some adults, formed an organization called Kids Against Pollution, and their cute and highly visible protests against fast-food packaging became the subject of feature stories in local newspapers and television newscasts everywhere. (The average elementary-school student is probably unaware, incidentally, that he or she throws away three-and-a-half ounces of *edible* food a day; looked at another way, the average elementary-school student every month throws away the equivalent by weight in edible food of 300 Big Mac foam clamshells.) Also getting into the act was the Pro-Environment Packaging Council, a group with ties to some in the paper industry. Paper stood to gain mightily from any misfortune that might befall polystyrene.

McDonald's, meanwhile, had been working with its polystyrene suppliers to set up a recycling program for polystyrene, and was planning to announce in the fall of 1990 that it had forged ties with existing polystyrene-recycling facilities, that new polystyrene-recycling facilities would soon be coming on line, and that the company would begin trying to recycle all of the polystyrene from its franchises nationwide. As with many kinds of recycling, the biggest challenge after finding a market is being able to collect homogenous quantities of a given commodity in great enough volume—something that McDonald's, with nine thousand outlets nationwide, was uniquely suited to do. Whether that, coupled with a switch to one of the alternative blowing agents that have rapidly displaced CFCs, would have quieted the critics will never be known. Neither came to pass. For reasons that remain unclear, McDonald's abruptly decided, in November of 1990, to abandon polystyrene completely, in favor of plastic-coated-paper "quilt wrap" for hamburgers (less bulky than foam, but recyclable only with great difficulty) and plastic-film-coated cardboard cups for coffee (the kind it had always used for sodas and milkshakes). In a bland statement of justification that seemed to suggest a desire on the part of McDonald's simply to wash its hands of the whole issue and be done with it, the company's president, Edward Rensi, said: "Although some scientific studies indicate foam packaging is environmentally sound, our customers just don't feel good about it."

The McDonald's decision was hailed almost universally as, in effect, a victory for source reduction, and McDonald's reaped a harvest of public good will. But if a victory for source reduction it was, the victory may have been an equivocal one. A few months after McDonald's made its announcement Martin B. Hocking, a chemist at the University of Victoria, in British Columbia, published a widely discussed article, titled "Paper Versus Polystyrene: A Complex Choice," in the journal *Science*. Hocking's aim was to compare the "environmental merit" of paper and polystyrene in packaging, and he did so by focusing on the paper and polystyrene used in the manufacture of single-use hot-drink cups. His conclusions contained some surprises. First, he wrote, the production of the paper for a paper hot-drink cup consumes as much in the form of hydrocarbons (oil and gas) as does the manufacture of a polystyrene cup (which is largely *made* of hydrocarbons). Moreover, the production process for the paper cup requires a great many more chemicals than does that for the polystyrene cup: for paper, 160 to 200 kilograms of chemicals per metric ton of wood pulp, versus about 33 kilograms per metric ton of polystyrene. Hocking went on:

> Because 6 times as much wood pulp as polystyrene is required to produce a cup, the paper cup consumes about 12 times as much steam, 36 times as much electricity, and twice as much cooling water as a polystyrene foam cup. About 580 times the volume of waste paper is produced for the pulp required for the paper cup as compared to the polystyrene requirement for the polyfoam cup. The contaminants present in the wastewater from pulping and bleaching operations are removed to a varying degree depending on site-specific details, but the residuals present in all categories except metal salts still amount to 10 to 100 times those present in the wastewater streams from polystyrene processing.

On a per-cup basis, Hocking found the air emissions from the production of polystyrene to be about 60 percent lower than those from the making of the paper for the hot-drink cup. Of course, Hocking observed, even if the polystyrene is blown up with pentane rather than with CFCs, there will be a negative effect on the ozone; but, he added, polystyrene's "contributions to ozone and as a 'greenhouse gas' are almost certainly less than those of the methane losses generated from post-use disposal of paper cups in landfill sites." (Well, to be fair, one must add this proviso: "if the paper cups biodegrade.") As for the recycling potential, Hocking rated polystyrene "easy" and paper cups "possible," cautioning that hot-drink paper cups are excluded from current recycling programs because they employ a "non-water-soluble hot melt or solvent-based adhesive" and also because they employ a wax or plastic-film coating. On top of all this, Hocking noted, on a per-cup basis polyfoam cups are cheaper than paper ones. Hocking's findings tended to corroborate those of a "cradle-to-grave" analysis by Franklin Associates of the environmental consequences of producing, using, and disposing of ten thousand polyfoam cups as compared to the consequences of producing, using, and disposing of ten thousand wax-coated paper cups. Soon after Hocking's *Science* article appeared the fast-food chain Hardee's announced that it was "definitely considering" switching from cardboard cups to foam cups.

One does not need to resolve conclusively whether Hocking is right on the issue of polyfoam versus paper. A paper-industry spokesperson and several scientists engaged Hocking vigorously in a subsequent issue of *Science;* Hocking, in response, conceded a few points but fundamentally stood his ground. Perhaps the matter is a wash. Perhaps paper is, in fact, somewhat more advantageous than polyfoam. The lesson to remember about this particular foray into source reduction is that after an enormous amount of activism by thousands of very concerned people over the course of more than a decade, and after the expenditure of an enormous sum of money by McDonald's to make the switch, and after all the considerable dislocations the new fast food–packaging regime has entailed for plastics recyclers and McDonald's suppliers—after all this, there would seem to be very little to show in terms of any real amelioration of the garbage situation. The famous phrase of the poet Horace about "laboring to bring forth a mouse" may very well apply to these circumstances.

Polyfoam totaled only some 4 percent by weight of all the garbage produced by McDonald's overall. Most of the garbage produced by the franchises—45 percent—was always plain old cardboard (for the packages that buns, patties, napkins, and bags come in, for example) and paper, just as in landfills; another 34 percent has consisted of various kinds of organic debris, such as egg shells, coffee grounds, and uneaten food. Now there will just be that much more paper.

To its credit, McDonald's did not terminate its garbage-management efforts with the polyfoam ban: In partnership with the Environmental Defense Fund, it is looking at ways it can more effectively cut down on the big-ticket waste it contributes to the municipal waste stream—recycling its cardboard boxes and composting its food debris. Those steps, at least, could actually make a difference.

Before leaving McDonald's behind, it may be worth mentioning a somewhat odd archaeological exercise conducted by the Garbage Project, one that involved a distant ancestor of the McDonald's clamshell—the ceramic Uruk bowl. The Uruk people flourished in what is now modern Iraq during the period 3300–3100 B.C.; their culture was named for the ancient city Uruk, which is today known as Warka, and lies in the desert halfway between Baghdad and Basra. The Uruk people are remembered for having given us several things, including the earliest samples of cuneiform writing that have survived. They also seem to have relied heavily on that bane of source-reducers, the food container that is used once and then thrown away. Uruk bowls, which are relatively small (they hold about half a liter) and are very simply made, have long intrigued archaeologists because they are found by the thousands, not scattered but concentrated in large groups, and are mostly intact—a rare combination of circumstances (see Figure 10-C). Ever since large numbers of these clay bowls were found during excavations at Nineveh, in 1929–32, archaeologists have tossed about theories among themselves over what the use of the bowls might have been. The fact that all of them were of cheap manufacture and yet were so often found in an unbroken state suggested strongly that they had not been used over a period of long duration, and the fact that they were so frequently

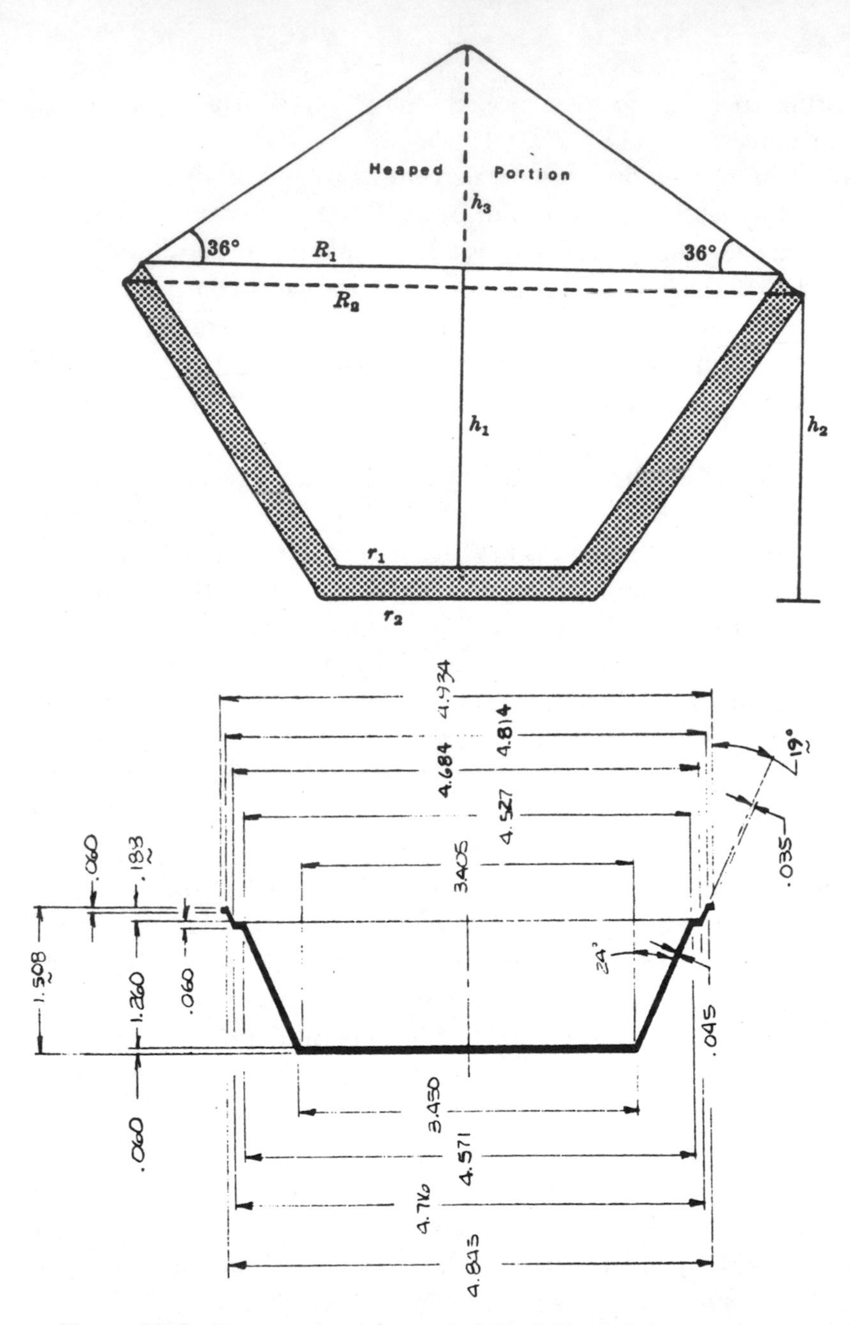

Figure **10-C.** Cross-section of a typical Uruk bowl; below it, for purposes of comparison, a detail from the original design specifications for the lower half of the McDonald's foam clamshell, reproduced to scale.

SOURCE: Thomas Wight Beale, "Bevelled Rim Bowls and Their Implications for Change and Economic Organization in the Later Fourth Millennium B.C.," *Journal of Near Eastern Studies,* October, 1978; McDonald's.

found in large groups strongly suggested the possibility of mass discard and single use, perhaps on some sort of special occasion. Most of the stabs at explanation have involved food. Perhaps, some think, they held votive offerings of food "to scare demons away from houses" (as Sir Max Mallowan believed). Perhaps, others have argued, they held offerings of food brought by the faithful on special occasions to temple personnel. And perhaps, still others contend, they were meant to hold standardized rations for laborers.

Because the information is pertinent to landfill issues, the Garbage Project has always been interested in the comparative densities of discarded materials. The question naturally arose one (idle) afternoon: What if fast-food restaurants had to use Uruk bowls as their single-use food containers? A Garbage Project staff member named Timothy Jones looked into the matter. Figure 10-D compares the solid-waste consequences of packaging one hundred hamburgers in Uruk bowls versus packing them in (now old-fashioned)

Figure 10-D. For purposes of comparison it was assumed that Uruk bowls can hold four hamburgers, whereas foam and paper clamshells can each hold one. The chart shows the garbage that would result (by both volume and weight) from packaging a hundred hamburgers according to each method.

SOURCE: Timothy Jones, The Garbage Project

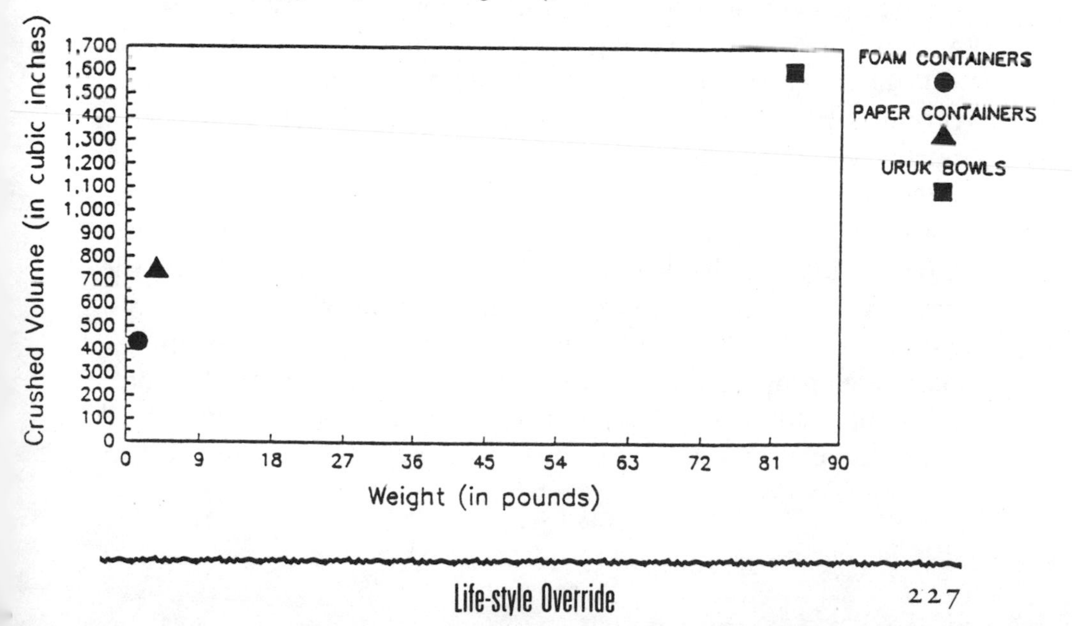

McDonald's foam clamshells. Those who worry about fast-food packaging clotting the solid-waste stream should be thankful that the fast-food era arrived when it did and not five thousand years sooner, when the single-use food container was in its infancy—that it arrived, one is tempted to say, in an age of Kroc and not of *the* crock.

Even if product bans and wholesale replacements in products of one type of material for another were capable of making a big difference in the solid-waste stream, they wouldn't necessarily be the most advisable source-reduction strategy. The solid-waste stream is not fed by many separate and independent pipelines, any one of which can be shut off without affecting the others. It is a vast, interconnected, impossibly complex system. Indeed, to think of the solid-waste stream as consisting of something called "garbage" is in a way naive. What it really consists of is the individual life-styles of 250 million interconnected Americans, and those 250 million interconnected lives resist micromanagement. The most radical advocates of source reduction hope that by opening a sluice gate here and closing one there and otherwise fiddling with the system's controls—in effect, fine-tuning the nation's garbage—they will bend the system to their will. It will not happen. Governments and referendums may force people to stop using this and stop using that, and there may be the odd individual success, but the full range of consequences will always be beyond predicting, will be enormous, and as often as not will interfere with entrenched life-styles and forms of convenience, breeding potent animosity. Along with "greens," pollsters are now turning up growing numbers of irritable "browns."

A more sensible approach to reducing the volume of the solid-waste stream—the goal of source reduction—would be to pay less attention to any individual source and more to the overall effect. The best way to do this is through the simple expedient of market forces: Making people who generate lots of garbage pay more for its disposal than people who don't generate very much garbage. This proposal may strike some as bizarre, because in 40 percent of all American cities (according to one recent survey) residents are currently not assessed any separate fees for garbage disposal at all, and in many of those cities that do charge for garbage disposal the fees are assessed at a flat rate. Neither regime represents a formula for reducing the volume of garbage. In contrast, the city of Seattle and some other municipalities provide their citizens with an economic incentive to reduce the volume of their solid-waste discards. In Seattle, residents subscribe to a garbage-collection program based on the number of 19-, 32-, or 60-gallon cans that they fill up each week; the monthly fee for each type of can, and for each additional can of whatever type, goes up exponentially with volume to discourage casual profligacy. At the same time, Seattle provides separate containers for recyclables and collects these recyclables without charge. In 1981, when Seattle inaugurated its system, the average household filled three-and-a-half 32-gallon garbage cans a week. The average today is a little more than one 32-gallon can a week.

Seattle's program works because it is uncomplicated and harnesses the power of financial incentives to promote source reduction and recycling. Rather than telling people precisely how their life-style must change—that is, rather than overriding their life-style, and overriding everyone's in exactly the same way—it holds out incentives for life-style changes and allows people to decide which changes, if any, they are going to make. It says, in other words, Here's the bottom line: You figure out how to get there. It acknowledges, in the end, that the social engineering of life-styles is fraught with obstacles. The diversity of life-styles in America tends to conceal the fact that, individually, life styles are surprisingly rigid things—rigid and, as a good deal of Garbage Project research shows, made up of some surprisingly perverse ways of doing business.

One of those perverse tendencies in behavior—which indirectly but powerfully supports the wisdom of garbage-limitation efforts such as the one undertaken in Seattle—has recently come to light as a result of Garbage Project sorting that has been conducted over the years in Phoenix and Tucson. The phenomenon in question can be thought of as a kind of Parkinson's Law of Garbage. The original law, which was formulated in 1957 by C. Northcote Parkinson, a British civil servant based in Singapore, states: "Work expands so as to fill the time available for its completion." The Parkinson's Law of

Garbage states: "Garbage expands so as to fill the receptacles available for its containment."

The evidence here is suggestive. During the past decade many municipalities have moved from a system in which homeowners provided their own garbage cans and sanitation workers emptied them by hand into trucks, to a system in which the city provides homeowners with special garbage containers that trucks can empty mechanically. The object is to save labor costs and reduce worker injuries. Of course, the city cannot make available a choice of a dozen or two dozen different sizes of garbage containers to its residents, calibrated according to household size; some mechanized trucks can handle only a limited number of sizes. And because large households must be accommodated, most city residents therefore receive a very large—90-gallon, in many cases—wheeled container.

In 1980 the city of Phoenix adopted such an automated system, with 90-gallon containers, and a subsequent Garbage Project study (in 1988), unrelated to the new containers, revealed that the per capita generation of garbage seemed to have become abnormally high, at least as compared to Tucson, which existed in a similar environment a mere 100 miles away. This was no illusion. A review of data for the years 1975 through 1990 showed that, although the amount of garbage collected per household per week had remained almost constant during this period, the number of people per household had in fact shrunk by some 25 percent. Nothing much more was made of this finding until Garbage Project researchers began analyzing data from areas of Tucson that had recently converted to a mechanized collection system similar to that of Phoenix—and realized that garbage-generation rates of sample households had shot up to a level markedly higher than long-familiar levels. Indeed, the increase amounted to about a third—comparable to the apparent increase in Phoenix.

Further investigation revealed that other cities that have gone the way of Tucson and Phoenix are also registering significant increases in per capita garbage generation. In Sacramento, the yearly per capita haul has risen from about 1.4 tons before mechanization to more than 1.8 afterward, an increase that has occurred even as tipping fees have more than doubled. So much for the savings in labor costs. In Dodge City, Kansas, a sanitation department official expressed surprise at the results of a pilot program in which people were given huge, 120-gallon garbage containers: "People filled the suckers up." In Beverly Hills, some neighborhoods have been given *300*-gallon containers, and one can only wonder what effect such encouragement will have on a community whose discard patterns are already excessive. Beverly Hills is the kind of place, according to sanitation officials there, where some homeowners regularly pick up the sod and throw out their entire lawns twice a year, to ensure that the type of grass will be appropriate both in summer and winter.

The dynamics of the Parkinson's Law phenomenon with respect to garbage are quite simple. When people have small garbage cans, many of the larger kinds of garbage that they have on their hands—certain kinds of yard debris, old cans of paint and containers of various chemicals, broken furniture and other damaged items perpetually awaiting repair, rusted tricycles, bags of old clothing—do not typically get thrown away. Rather, these "outlier" items, some of which are examples of what is known to archaeologists as "provisional refuse," gather dust in outdoor junk piles and in basements, attics, and garages, often until a residence changes hands. But when homeowners are provided with these plastic mini-dumpsters—these lidded and cavernous garbage mausoleums—they are presented with a new option. Before long what was once an instinctive "I'll just stick this in the cellar" or "I'll just throw this behind the garage" becomes an equally instinctive "We can get rid of this" and "I'll bet this will fit in the dumpster."

The Garbage Project has compared the contents of Tucson garbage collected from the same neighborhoods before and after the 90-gallon containers were adopted, and the results confirm the explanation just given (see Figure 10-E). The increase in solid-waste discards was substantial: from an average of less than fourteen pounds per biweekly pickup to an average of more than twenty-three pounds. The largest increase was in the yard waste category, followed by "other" (all those broken odds-and-ends), food waste, newspapers, and textiles. The first pick-up of the week was consistently heavier than the second one, reflecting the accomplishment of weekend chores, and the discards in that pick-up contained larger amounts of hazardous waste than the Garbage Project had come to expect in a typical load of garbage. That consequence of the intro-

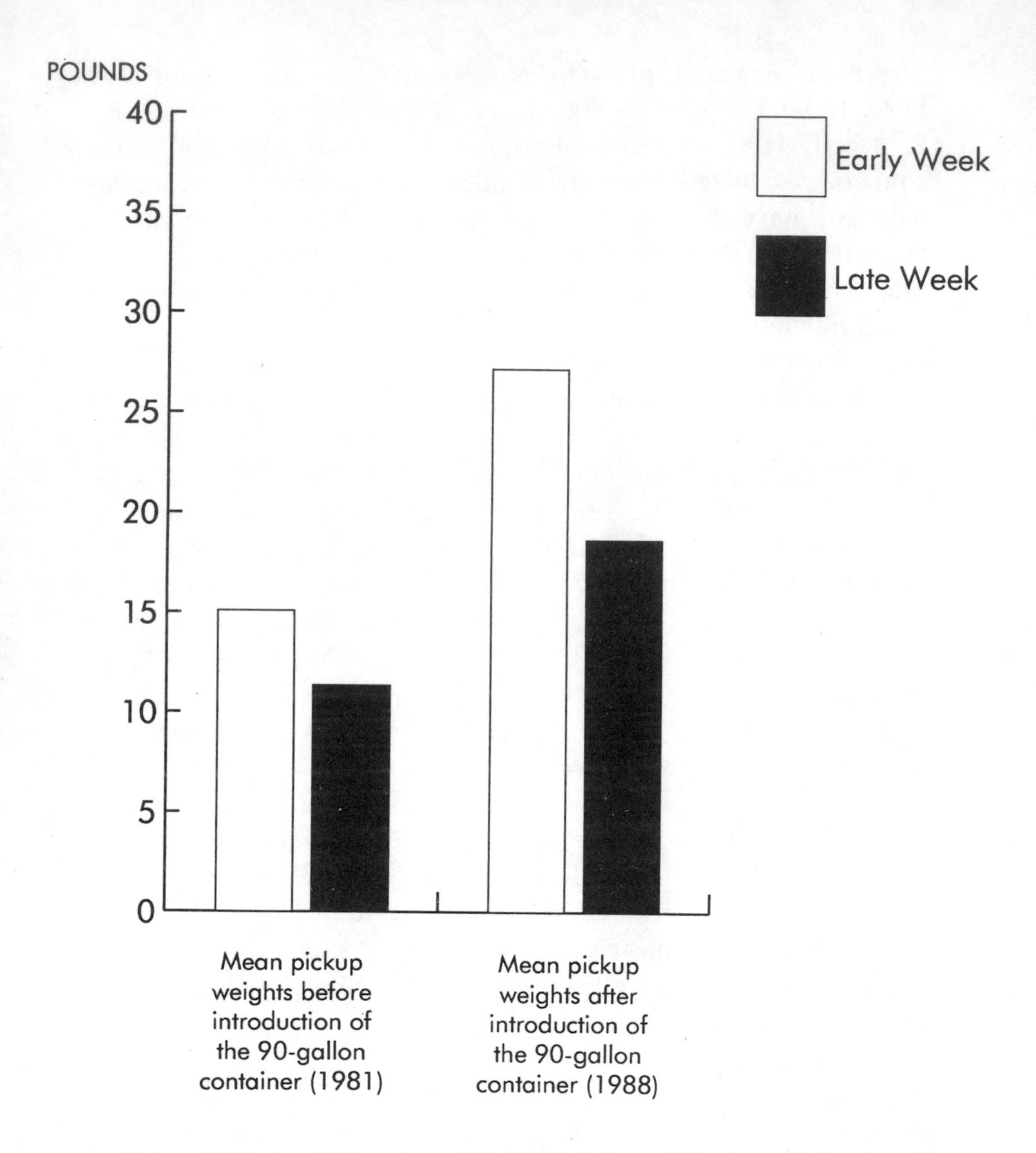

Figure **10-E.** A manifestation of Parkinson's Law of Garbage. Following the citywide introduction of oversized garbage cans in Tucson the size of the average pickup suddenly increased.

SOURCE: The Garbage Project

duction of 90-gallon containers, if nothing else, should be a matter of special concern.

Thus, once again, the American household has demonstrated its unwitting capacity to subvert public purpose. When it comes to managing garbage, the pitfalls that may hamper effective policies are not only legion but also, frequently, hidden. Are there ways to avoid such pitfalls, or at least keep them to a minimum? Yes, possibly there are. We'll discuss some of them in the next chapter.

CHAPTER 11

THE TEN COMMANDMENTS

The New England Resource Recovery Conference and Exposition meets every year in late spring to confront a wide variety of issues involving garbage, and it affords a convenient platform from which to view the diverse and fractious world of garbage in microcosm. The sponsors range from agencies of the government (the U.S. Environmental Protection Agency, the Council of Northeastern Governors) to nonprofit groups (the New Hampshire Resource Recovery Association) to industry groups (the Aseptic Packaging Council) to individual corporations (Procter & Gamble; Johnson Wax). The companies maintaining booths in the vast exhibit hall have names like Earthgro Compost Services and Global Ozone Solutions and Enviro-Fibre, Inc. The two giants of the garbage-services industry, Browning-Ferris Industries and Waste Management, Inc., are of course there in force, their representatives explaining that, yes, their companies had wide experience in virtually any garbage-disposal regime a community might want to implement, be it recycling, incineration, composting, or landfilling, or an "integrated" approach that combines all four in some fashion, tailored to local circumstances. In addition to the exhibits some forty or more seminars and workshops are typically offered at the conference, the topics leaving no aspect of solid-waste management unaddressed. At the 1991 session the seminars had titles like these:

"Where Do All the Tires Go?"
"New Opportunities for Reducing and Recycling Medical Waste"
"Ash Stabilization and Reutilization"
"Construction and Demolition: The Neglected Challenge of the '90s"
"Beyond NIMBY"
"Maintaining Composting Quality"
"Entering the New Age of Ordinances, Fines, and Penalties"
"Processing and Marketing Bagged Recyclables"
"Establishing Ongoing Household-Hazardous-Waste–Collection Programs"

The conference was held in the city of Springfield, Massachusetts, and those who attended could find relief from its business by signing up for tours of such places as Springfield's sixty-eight-acre double-lined landfill and the Wheelabrator waste-to-energy facility in nearby Millbury.

At the conference there is a lot of big equipment on display, such as portable trommels and tub grinders, and the men on hand to explain it tend to have names on their shirts, and are not afraid to smoke. Elsewhere in the exposition hall there are federal bureaucrats who dress and speak carefully; recycling advocates of various kinds, from pallid former hippies to clean-cut outdoorsmen to jut-jawed entrepreneurs; tough-talking scrap dealers, with their big hands and relentless voices; a smattering of corporate executives, well-educated, -heeled, and -tanned; and a good many officials of local governments, all of them with some down-to-earth problems, a need for help, and not very much money. This is a group of people who speak a common language. They all know what is meant by such phrases as "selective catalytic reduction" and "listed waste." But their interests may vary widely, as may their conceptions of the best way to dispose of municipal solid waste. Though bound together by their focus on a common problem, they do not by any means speak or act as one.

A major fault line became apparent at the 1991 conference when Barry Commoner addressed a lunchtime crowd and repeated his familiar claim that some 80 to 90 percent of all household waste is potentially recyclable—a figure based on a pilot program conducted for ten weeks among 100 volunteer households in East Hampton, New York. (East Hampton has since inaugurated a recycling program based on a blueprint provided by Commoner's Center for the Biology of Natural Systems.) As Commoner spoke, roughly half the audience listened attentively, held by his words and his vision. But other people shifted impatiently, and a number of them drifted away from their tables to form a knot at the rear of the auditorium. They were all affiliated with various waste-to-energy facilities, and there is no love lost between many such people and those who most ardently promote recycling programs, both groups sometimes competing to get their hands on a single community's garbage. "Can you believe this?" one of the waste-to-energy people, referring to Commoner's remarks, whispered to a companion. Everyone around him laughed.

The points of friction among those who handle America's garbage derive in part, of course, from the precarious economics of garbage disposal. They also derive in part from the fact that differences of opinion are colored (and complicated) by the presence of conflicting corporate and environmentalist ideologies. And they derive as well from the tendency among any one set of players to have less than a firm grasp on the social, technical, economic, and political realities that shape the thinking and the approach of any other set of players. Perhaps more than anything else, then, what we need as we confront the garbage problems that do exist in America is a sense of perspective.

If we step back and take the long view, several things immediately stand out. One involves America's place in the historical scheme of things. For all our technological sophistication, Americans are not that different from those who inhabited most of the world's other great (and ostentatious) civilizations. Our social history fits rather neatly into the broader cycles of rise and decline that other peoples have experienced before us. Over time, grand civilizations seem to have moved from efficient scavenging to conspicuous consumption and then back again to the scavenger's efficiency. It is a common story, usually driven by economic realities.

In their beginnings most civilizations, ancient and modern, have made efficient use of resources. The Early Preclassic Maya, who inhabited the rainforests of the southern Yucatan between 1200 and 300 B.C., seem to have lived relatively simple farming lives. They built a few small temples, constructed large houses out of thatch on low dirt platforms, and interred their dead with one or two monochrome pots. Subsequently, something extraordinary happened. The Maya, entering their Classic phase around A.D. 300 (archaeologists use the term "classic" to refer to societies at the zenith of their wealth and power), acquired a taste for excess: fancy ceremonial clothes and feathered headdresses; tall temples with intricately carved façades; commanding stelae sculpted with figures of haughty rulers in full regalia; and lavish burials with grave offerings of shell mosaics, carved jade, large, eccentric flints, and richly painted polychrome pots. This Maya cult of conspicuous consumption spread throughout southern Mesoamerica. Then, during what is known as the Decadent Maya period, temples became small again, tombs were reused, and burial offerings contained only a few pieces of broken pottery or chipped and shattered obsidian knives. Whatever the stimulus—and most explanations implicate some sort of vast economic erosion—in what archaeologists call "decadence" we today can see efficient resource utilization. Among the Decadent Maya everything was recycled and reused, and virtually no resources were put away beyond easy retrieval. The Decadent Maya were living on the edge; they probably had no choice. And in the end, for all the recycling, reuse, and conservation, even the rather pathetic material ostentation of the Decadent Maya disappeared.

The United States today remains well within a Classic phase. Our waste-generating habits are robust, and they would be robust even if the often conflicting dreams of the source reducers and the recyclers and the waste-to-energy advocates were somehow all, by a miracle, to come to pass.

For all the aggravation of its consequences, waste generation of great magnitude has historically been a sign of economic and social vitality. The question is: Can we put that vitality to good purpose? No society in history has been better equipped to deal sensibly with large amounts of garbage than ours is, and owing to the legitimate concerns of environmentally minded people in and out of business and government the means at our disposal for handling garbage are improving all the time. While our knowledge of the solid-waste stream remains relatively primitive, we know vastly more now than was known a mere two decades ago. Perhaps for the first time since human beings left their hunter-gatherer life-style behind them, it is now possible to imagine a truly rational garbage regime: one in which we maintain the core character of the way of life but at the same time take steps voluntarily to adopt "decadent" behavior. The point is to take those steps to conserve, reuse, and recycle on our own terms, long before time and resources and society's margin of error run out, and we are forced to take the necessary steps under conditions far less favorable. As we contemplate the prospect of Decadence Now, a handful of admonitions—the ten commandments, as it were—should be borne in mind.

1. *Don't think of our garbage problems in terms of a crisis.*

The garbage problems that Americans face are real, and solving them in sensible ways will cost money and make demands on our life-styles. But it will not cost all that much money or make all that many demands. In some respects the key is to remain calm. Our garbage is not about to overwhelm us; there are a number of options available; and most communities have time to think about those options and choose among them wisely. The worst thing to do would be to blow the problem out of proportion, as if garbage were some meteor hurtling toward the planet. The term "crisis" seems to demand immediate, drastic action, and this sort of action is indeed often what one hears proposed. But crisis thinking typically results in ill-conceived and counterproductive initiatives. A more rational garbage policy would consist of muddling along, making improvements at the margin all the time, applying the fruits of advancing technology and of new knowledge about human behavior, thinking through the second-, third-, and fourth-order consequences of proposed initiatives—and then turning our minds to other things. There is, after all, a country to run.

2. *Don't bow before false panaceas.*

Americans seem to expect that there ought to be one simple "silver bullet" that will enable us to get rid of all our garbage efficiently and safely, and maybe even make money on the side. No such silver bullet exists, although the most committed partisans of each of the main solid-waste-management strategies may insist that *their* approach comes close to being one.

Each approach—source reduction, recycling, incineration, and landfilling—has real advantages, but each also comes with significant disadvantages. Source reduction doesn't diminish the volume of garbage by all that much, and it eats away at the conveniences that lie at the heart of our life-styles. Recycling is a fragile and complicated piece of economic and social machinery—a space shuttle rather than a tractor; it may break down frequently. The manufacturing portion of the recycling loop also results in the production of pollution. Incineration causes some amount of hazardous waste to be emitted through smokestacks, and it is virtually impossible to quantify what the actual health risks may be from various levels of emissions; also, a byproduct of incineration is some quantity of toxic ash, which must somehow be disposed of. Landfills, for their part, take up a considerable amount of space; some discharge toxic leachate; and the land can never be returned to a truly pristine state.

The goal must be to discern the varying roles that each of these approaches should play locally in America's widely disparate communities and regions. Extremism with respect to garbage solutions is ideologically satisfying, and some of the nation's most prominent extremists on garbage issues have played a valuable educational role. In the real world, however, an insistence on utopia always causes trouble.

3. *Be willing to pay for garbage disposal.*

The misdirected search for a garbage panacea typically involves a search not only for an appropriate disposal method but also for a method that is cheap or, better, one that holds out the possibility of turning a profit; a disposal method that, additionally, holds out the promise of conserving resources, creating energy, and saving the world. If a certain disposal technology happens to kill two birds with one stone, as waste-to-energy plants do, then that is fine. But our focus should be primarily on the first objective: getting rid of garbage. It is enough to do that safely and well, and if there are ways to do so, they are worth paying for. Though rarely thought of in this way, garbage disposal is no less an essential public service than is police and fire protection or sewage treatment, and we should be just as willing to pay full value for it. "Zero-net-cost" provisions in municipal recycling contracts—that is, stipulations that a community will not pay more per ton of garbage to a recycler than it was paying under a previous disposal regime—fail to take into account such ancillary benefits of recycling as resource conservation and the saving of landfill space. In the management of garbage disposal, adopting cheap solutions is usually a prelude to encountering expensive surprises. Shortcuts will always turn out to be the long way around the problem.

4. *Use money as a behavioral incentive.*

A corollary of commandment number 3 is a general point made early on, one that cannot be overemphasized: Desirable things happen to garbage when someone stands to make (or save) money as a result. This statement is valid for every garbage operation and for all garbage-producing and garbage-handling units. While the behavior of masses of people with respect to garbage is frequently perverse, as we have seen, the presence of money as a behavioral force has a pronounced rationalizing effect. This conclusion is hardly revolutionary—after all, it underlies the operations of a market economy. But it has rarely been at the forefront of public thinking about garbage, and it is rarely reflected in the policies of local communities. For example, as the case of Seattle demonstrates, charging people on a progressive scale for the amount of nonrecyclable garbage that they put out for disposal, while at the same time not charging anything at all for recyclables, has precisely the effect that economic theory would predict: Recycling rates improve, and the overall volume of nonrecyclable garbage left out for disposal diminishes. And yet, as noted, despite this proven relationship, most communities pay it no heed. It is difficult to think of an area of solid-waste-management where admonitions to virtuous collective behavior would prove to be a more effective spur than would some simple tinkering with fees. The impact would ripple backward, providing incentives for manufacturers to use less packaging and to design products for ease of recycling.

5. *Distrust symbolic targets.*

If the United States had a solid-waste czar and a genie granted him one wish, the wish might very well be to have put at his disposal all the time, money, and energy that has been wasted on attacking categories of garbage that have high public visibility but in a larger sense matter very little if at all. Foam containers, fast-food packaging, disposable diapers, and certain kinds of plastic fall into this category. Everyone takes a visceral or aesthetic dislike to at least some of these things, but dislike is an imperfect guide to the crafting of intelligent public policy. The fact is, if these items disappeared tomorrow we would still have to answer "No" to the question: Have any of our fundamental garbage-disposal problems been addressed?

It is the nature of potent symbols that they are difficult to resist, but a link between the symbol and what is symbolized is sometimes tenuous. This seems the case with garbage to an unusually great degree. There may be no more vivid a symbol of the garbage crisis than *Mobro 4000,* the garbage barge, and yet the garbage barge never did represent what it has by now come to stand for. The garbage barge was not part of some last-ditch attempt by a desperate community to find someplace—anyplace—to put its mounting piles of garbage. It stemmed, rather, from miscalculation: An entrepreneur believed that he could make a fast profit by agreeing to accept the garbage of the town of Islip—and along with it a sizeable tipping fee—and then quickly offloading the garbage in some benighted locale that would consider itself lucky to receive a much smaller tipping fee. The gamble didn't pay off. *Mobro 4000* ought to be remembered as a symbol of greed, or perhaps bad luck. But as we all know, it isn't.

6. *Focus on the big-ticket items.*

This is the necessary complement to "Distrust symbolic targets." Taken together, two kinds of garbage—paper and construction-and-demolition debris—account for well over half of America's general refuse. Oddly, they receive far less than their proper share of the publicity attendant upon America's garbage problems. And yet, were some way to be found to radically reduce the volume of material that must be disposed of in these two categories of garbage—whether through recycling or other means—one could answer with an emphatic "Yes" the question posed just above, for some of our fundamental garbage problems *would* have been addressed. Archaeologists of the future digging into landfills in the Toronto area—as the Garbage Project has just done—will discover that the volume of newspapers declined by 66 percent in the course of the 1980s: the consequence of dogged commitment to a recycling program that did not come cheap. The wood and concrete components of construction-and-demolition debris are also recycled in the Toronto region—a feasible undertaking if the two categories are kept separate at construction sites.

7. *Buy recycled and recyclable products.*

The various kinds of collection and sorting programs all around the country for various kinds of recyclable garbage seem to be doing fine. Consumers could give recycling a big boost—could help close the recycling loop—by making a point of buying goods and packaging with a high recycled content. The effect would be to stiffen demand where currently it may be soft, and it would send a message to manufacturers that they would be likely to heed. The city of Phoenix recently launched a citywide recycling program, and its director, Jack Friedline, has gone out of his way to explain to Phoenicians that buying recycled products and products whose packages can be recycled—and, especially, buying those kinds of recyclables that have the most resale value (for example, aluminum cans versus glass bottles)—may provide the economic lift that a workable recycling program needs. It may also further encourage industry to design products with ease of recycling in mind.

For consumers, buying recycled products would represent a minimal investment of time and energy. But consumers will have to become garbage literate, because labels can be deceptive. For example, the word "recycled" on a package generally means not that a product has been made, at least in part, out of something that a consumer once bought and then turned in for recycling, but rather that it has been made in part with scrap left over from the normal manufacturing process—business as usual in any well-run factory. The label one needs to look for is "post-consumer recycled," and ideally the label will include a percentage, as in "30 percent post-consumer recycled." Anything over 10 percent is worthwhile. Consumers also need to know the pertinent legends and markings on labels, which indicate that various paper and plastic items are major players in the recycling loop. (New organizations such as Green Seal may increasingly be able to provide guidance on this and related matters.) Individuals, by the way, aren't the only consumers who can help sustain recycling by buying recycled products; so can governments, and so can businesses.

8. *Encourage modest changes in household behavior.*

Those just-mentioned admonitions to virtuous collective action may be an inefficient tool of public policy, but they have a role to play. This is especially the case at the most basic social level, that of the household, where a few relatively simple changes in standard operating procedure could, over time, have a beneficial effect on the overall solid-waste situation. Although it makes no sense for governments to promulgate a wide and complex array of suggested household reforms in this regard—which would no doubt be as confusing as the surfeit of advice about health and diet has now become—they could usefully hammer away on two or three fronts.

One is food waste. Between one-tenth and one-quarter of all the edible food a family buys gets thrown away; this does not include food-preparation debris, such as rinds, peels, skins, and so on. No one who wastes food feels good about doing so, and reducing food waste makes moral and economic sense. (Remember, by the way, the First Principle of Food Waste. The more repetitive your diet, the less food you waste.) A second issue is composting. Almost any property capable of producing yard waste is also capable of lodging a compost pile. Some localities have banned yard waste from landfills, and such moves make sense. Composting is not a difficult procedure. Nature does most of the work, and the end result is a useful product. A third issue is hazardous waste. To a significant degree, the level of toxicity of the leachate in a community's landfill is the end result of thousands of individual decisions by local households. Those decisions could be a lot more intelligent: Use hazardous products as they are intended to be used; use them up before throwing them away; participate in local hazardous-waste-collection programs.

Getting people to change behavior in the ways outlined above will have no immediately obvious positive consequences. The analogy here might be to the antismoking campaign, where sustained attention to the problem eventually resulted in behavioral change that has occurred at a relatively slow pace and whose beneficial consequences are as yet difficult to apprehend in the statistics. But no reasonable person doubts the wisdom of the strategy.

9. *Be reasonable about risk.*

The garbage debate turns on two issues, volume and risk, and of the two the discussion of risk is the more emotional. To some degree, that is as it should be. Each of the approaches that Americans take to the disposing of garbage entails some modest measure of compromise with respect to current or future public health. And, obviously, we need to ensure the integrity of our disposal systems. The linings and caps of our sanitary landfills must remain unbreached, the emissions from our incinerators must be kept as low as possible, and the toxic byproducts of our recycling mills must be disposed of with great care.

That said, it may be time to calm down a little. There is a concept in theology known as a "scrupulous conscience." The term applies to a person whose sense of personal transgression against the divine is so exquisite—whose sensors of sin amplify the most unthinking of technical violations into drumbeats of damnation—that normal life collapses into incapacity and self-loathing. Beginning in the late 1960s and early 1970s, the United States gradually acquired (and, indeed, cultivated) a scrupulous conscience with respect to risk—a point made by several prominent contributors to a recent special issue on risk of the scholarly journal *Daedalus* (Fall, 1990). The point here is not that an environmental conscience wasn't needed—it obviously was, as was the regulatory outburst that occurred in response. It is still needed, as is regulation. But we must guard against turning into a society that regards all risk as unacceptable, even when accepting some small risk may alter a relatively more odious status quo.

10. *Educate the next generation—without the myths.*

The present generation of adults, together with its immediate predecessors, is responsible for the garbage problems that we have. The next generation will inevitably come to share in that blame unless some sort of educational intervention dispels the major misperceptions. Those misperceptions—about paper and plastic, about biodegradation, about recycling, about virtually every aspect of garbage management—have been pointed out throughout this book. (The misperceptions are also embodied in countless well-meaning school programs about garbage and the environment.) Along with the lessons their parents are learning from recent scholarship and hard-knocks experience, the young need to be inculcated with a sense of proportion. Garbage has an undeniably large symbolic presence. It lacks the tidiness of being merely itself, representing as it does the back end of our life-styles. And, yes, without sustained attention garbage problems can certainly get out of hand. But once reasonable policies are in place, the task of disposing of garbage should be neither Herculean nor hideously complex.

A final point: The garbage problems that the United States has experienced will have had an unexpectedly welcome outcome if they drive home a lesson that is relevant to a broad spectrum of public-policy issues: namely, that mental realities and material realities are not congruent, and in many cases do not even closely approximate one another. This conclusion has emerged time and again from Garbage Project studies, and there is a growing body of anthropological literature to substantiate one's common-sensical hunch: that the phenomenon is a general one, cropping up in all fields of endeavor. Disdained commodity though it be, garbage offers a useful, if ironic, reminder of one of the fundamentals of critical self-knowledge—that we do not necessarily know many things that we think we know. That is not the usual starting point of most discussions in America, especially political ones. But it is not a bad starting point at all.

ACKNOWLEDGMENTS

Many people have contributed to this book, and in a variety of ways. First among them are the midwives who assisted at the birth of the Garbage Project—Ray Thompson, Tom Price, and Swede Johnson, to begin with, and also A. Richard Kassander, Jr., Fred Gorman, and Mark Suckling.

For two decades a small army of solid-waste and operations specialists have sustained the Garbage Project's research by providing garbage for analysis. They include, in Tucson: Sonny Valencia, Veronica Sainz, Hector Loya, Ron Meyerson, Yiki Martinez, Ray Murray, George Hall, Chuck Sheffield, Gilbert Mejias, Waldo Urbano, Oscar White, Ralph Quihuis, Leon Ellis, Frank French, and Edward Sandoval. In Phoenix: Ron Jensen, Jack Friedline, Gene Gabrielli, Tom Webb, and Wanda Wildman. In Tempe: Ron Otwell and Robert Hughes. In Milwaukee: Herb Goetsch, Ignatius Balistrieri, and Raymond Caplan. In New Orleans: Patrick Koloski and Bert Klienpeter. In the San Francisco Bay Area: Joe Garbarino, Guido Zanotti, Julio Dami, Tony Galli, Bob Biasoti, and Valerie Lenz. In the Chicago Area: Louie Boulander, Jerry Hartwig, and Denny Urbanski. In New York City: Phil Gleason, Robin Geller, Bill Young, and Mahesh Desai. In Collier County, Florida: Bob Fahey.

A number of people have contributed significant research questions and research funding to the Garbage Project. Their ranks include Edward H. Bryan (National Science Foundation), Haynes Goddard and Geri Dorian (Environmental Protection Agency), Betty Peterkin (U.S. Department of Agriculture), Gladys Block (National Cancer Institute), Richard Weichman (American Paper Institute), Burdette Bridenstein (National Live Stock and Meat Board), Loraine Russell Jackson (Association of Bay Area Governments), and the Centro de Ecodesarrollo and the Instituto del Consumidor, in Mexico.

Academic colleagues from many different fields have collaborated with the Garbage Project: Barry Thompson (University of Wisconsin, Milwaukee); Patrick Murphy (Marquette University); Dave Phillips, Jr. (University of California, Santa Cruz); Steve Cassels and Rolf Myhrman (Judson College); Ivan Restrepo (CECODES); Robert Ham (University of Wisconsin, Madison); Joseph Suflita, K. Gurijala, Melanie Mormile, and Frank Concannon (University of Oklahoma); Anna Palmisano, D. Maruscik, and Bernie Schwab (Environmental Laboratories, Procter & Gamble); Joe Robinson (The Upjohn Company); Nancy White and J. Rose (University of South Florida); Gail Harrison, Cheryl Ritenbaugh, Mike Reilly, Woody Bryant, and Chuck Gerba (University of Arizona); Jim Deetz (University of California, Berkeley); E. Wyllys Andrews and Kathe Trujillo (Tulane University); and Wendy Ashmore and Carmel Schrire (Rutgers University).

Also generous with advice and guidance have been colleagues and friends in the garbage field: Allen Hershkowitz, Denis Hayes, Dana Duxbury, Bill Brown, Terry Serrie, Chaz Miller, Leslie Legg, Jerry Hayes, Judd H. Alexander, Joan Lionetti, Bob Hamp, Ken Wills, Susie Harpham, Bob Hunt, J. Winston Porter, Iraj Zandi, Lewis Irwin, and Bill and Marge Franklin.

The ideas, fieldwork, lab and computer analysis, writings, and personalities of the current Garbage Project staff lie at the heart of the Project's contributions and lore. Doug Wilson, Masa Tani, and Susan Dobyns have each been key players for more than six years, as graduate students and then as colleagues. Tim Jones, Gerardo Bernache, Ramón Gomez, and Barbara Teso are following their lead. Other Garbage Project staff members have included Bruce Douglas,

Amy Foxx, Paul Friedel, Don Kunkle, Kim Murnbower, Randy McGuire, Sherry Jernigan, Kelly Allen, Bryan Johnstone, Mona McGuire, John Kerr, Don Grissom, Karl Reinhardt, Sherry McFate, Justine Shaw, Laura Schuchardt, Clyde Feldman, Mike McCarthy, and Marty Senour. Kathy Owen has kept the Garbage office functioning and landfill teams organized in the field. And to Doris Sample's skill and patience, a special thanks.

As every Garbage Project insider knows, Wilson Hughes has trained all of the Project's workers and has been there day in and day out to see to it that the Project's garbage gets sorted. Much of what has kept Wilson and the Project going is the esprit de corps that the volunteer student sorters have consistently brought to the garbage-sorting yard.

A word of gratitude is in order, finally, for those who have provided editorial guidance: Rick Kot, Corby Kummer, Raphael Sagalyn, Charles Trueheart, Lowell Weiss, and William Whitworth. Ethan Seidman, now at *Garbage* magazine, was an invaluable fact-checker, researcher, and all-purpose sherpa. And Anna Marie Murphy contributed not only her fabled good sense but also ample amounts of tolerance. Thank you.

The work of the Garbage Project has over the years been generously supported by a variety of government agencies, nonprofit institutions, and private companies.

GOVERNMENT

- National Science Foundation, Program for Research Applied to National Needs, 1975–76
- Environmental Protection Agency, Solid Waste Research Branch, 1978–79
- National Science Foundation, Program for Applied Science and Research Applications, 1977–78
- Environmental Protection Agency, Solid Waste Research Branch, 1978–79
- The New South Wales State Pollution Control Commission, Australia, 1978
- United States Department of Agriculture, Consumer Nutrition Division, 1980–82
- Instituto Nacional del Consumidor, Mexico, 1980–81
- National Cancer Institute, Department of Health and Human Services, Small Business Innovation Program, 1985

Environmental Protection Agency, Office of Solid Waste and Emergency Response, 1986–87
The Association of Bay Area Governments, California, 1986
Bureau of the Census, Center for Survey Methods Research, 1986–87
CECODES (Centro de Ecodesarrollo), Mexico, 1987–89
National Science Foundation, Environmental and Water Quality Engineering, 1986–89
The City of Phoenix, Arizona, 1987–88

RESEARCH FOUNDATIONS
Australian Museum Society, 1978
The San Francisco Foundation, 1986
Queens College Foundation, 1986

UNIVERSITY OF ARIZONA
The University of Arizona Foundation, 1978
The University of Arizona Foundation, 1983–84
Governing Committee of Social and Behavioral Sciences Research, 1985
The University of Arizona Foundation, 1987
The University of Arizona Foundation, 1988–89

PRIVATE INDUSTRY
Alcoa Aluminum, 1973
Chevron Oil, 1973–74
General Mills, 1973
Hoffman-LaRoche, 1973
SCA Services, 1973
Frito-Lay, 1976
American Can, 1976
Continental Can, 1977
American Can, 1978
Continental Can, 1978
Unilever-Australia, 1978
Solid Waste Council of the American Paper Institute, 1978–79
Gurnham and Associates, 1979
Solid Waste Council of the American Paper Institute, 1983–84
National Live Stock and Meat Board, 1984

National Live Stock and Meat Board, 1985
Amigos, 1985
Dames and Moore, 1985
Miller Brewing Company, 1985
California Beef Council, 1985
Castle and Cook, 1985
Heinz, 1986
Oscar Meyer, 1986
NutraSweet, 1986
Procter & Gamble, 1989
Franklin Associates, 1989–90